A Year Unplugged

A Family's Life Without Technology

Sharael Kolberg

Cover design by www.ebooklaunch.com

Photo by Blue Sky Studios

ISBN-10: 0988961040
ISBN-13: 978-0-9889610-4-3

For Jeff and Katelyn

Author's Note:

All of the events in this book are true. However, on occasion the day on which they occurred has been changed to make the story flow more smoothly. In most cases, names and/or identifying details have been changed to protect the privacy of those involved. Statistical information is from 2009-2010, the year in which this book took place.

Table of Contents

Introduction

I toss and turn in bed. The silence of the night is deafening. My sheets are damp with perspiration as my mind races with anxiety like a kindergartener dreading the first day of school. I keep repeating the terrifying phrase, "No Internet access." *How am I going to find out what the weather will be like to plan our family weekend outings, or get directions for a play date, or see what's playing at the movies for my monthly date night with my husband?*

And no TV. *What the hell am I going to do on Tuesday and Wednesday night without American Idol? Does that mean I actually have to sit and talk to my husband? What will I say?* Since our daughter was born five years ago, our lives have gone from fairytale romance to me looking and feeling like something out of Michael Jackson's "Thriller" video due to many months of sleep deprivation. And sex, hmmm, I think I vaguely remember what that is. *Is that clothing optional?* The thought of baring my post-baby body has kept me in baggy pajamas for an extended length of time.

And what's a mom to do without with Facebook for advice on the ever challenging issue of getting my daughter to "try just one pea" without her collapsing on the floor screaming, "Nooooooo," as if I were asking her to eat the poison apple from *Snow White*. And what if I miss an e-mail from one of my editors offering me the chance to write the story of a lifetime? Just what if *National Geographic* happens to think one of my witty ideas is worthy of their publication (I can dream)? Am I willing to risk missing out on the opportunity by not having access to my e-mail?

As tough as it may be, this is going to be our life for the next year as we embark on a journey to bring closeness to our family by distancing ourselves from our beloved technology. Can we do it?

The Birth of an Idea

Late one night, after Jeff tucked Katelyn into bed, he gingerly tip-toed out of her room looking less like his hunky-husband self and more like Sleepy, one of Snow White's dwarfs.

He plunked himself down on the couch next to me. I felt his warm skin next to mine and leaned my head against his chest. I cherished these brief moments of closeness in a life that always seemed to be on fast forward. We were tired.

Katelyn, five years old, had been sick for the past week and had no energy to do anything but watch back-to-back reruns of *Hannah*

Montana. It was easy to give in to her request to "watch just one more show" since I had dishes to do, laundry to wash, dinner to cook and work to catch up on. I felt a twinge of guilt using TV as a babysitter, but felt I had no other choice.

As Jeff and I sat in silence, I looked up at him and said, "Now that Katelyn's feeling better, we need to cut back on TV. I just don't think it's a good idea to let her watch so much of it."

"Why don't we try doing a week without TV?" Jeff said.

I shuddered at the thought of spending an entire afternoon trying to entertain Katelyn, an only child. What would we do for the four hours that needed to be filled between the end of school and dinnertime? Easy for Jeff to suggest giving up TV, he's not the one having to play hopscotch over and over, reading the same books every day, making animals from Play-doh, or playing "cashier" for the umpteenth time with a pretend cash register and plastic food.

It's not that I don't adore my daughter. She is the light of my life. But sometimes mommy needs a bit of a break from hours of *Romper Room*-type activities.

And how would Jeff and I spend our evenings without television to mask the truth that we had forgotten what it was like to have a passionate marriage? We had become so accustomed to turning on the TV after Katelyn was fast asleep. Rarely was there time set aside to just talk...or, heaven forbid, indulge in a lip-locked make-out session like frisky high schoolers. Over dinner, we would catch up on the day's events, but when was the last time we just sat and talked? I soon began to realize how dependent we had become on television...and all of our technology gadgets. We had gotten into the habit of spending our evenings with the TV on while simultaneously checking e-mail or seeing what our Facebook "friends" were up to.

I missed my husband. I missed the lively real-life chats we once had. I missed cozying up to a fire with a glass of red wine and just getting to know each other better.

So, I bravely said, "Why don't we do away with all technology?"

Jeff looked at me in disbelief and shook his head, "Why?"

"Why not?" I said. "It would be nice to have some fun family time in the evenings instead of turning on the TV and our computers every night."

He wasn't convinced. "I'm just so tired after working all day. By the time we get Katelyn fed, bathed and in bed, I just want to unwind," he said.

And this is why *Men are from Mars, Women are from Venus* is such a popular book. Men and women think differently and have different needs. I felt like a lawyer arguing my point. Family vs. Technology. After more discussion (and the threat of not having a home-cooked meal on the table when he gets home from work every night, not having his laundry washed, dried and put away for him, not having the cupboards miraculously always stocked with food, and not having floors so clean you could eat off them), Jeff somehow began to realize that it might be worth exploring life without technology.

"I don't think it will make that much of a difference to give up our technology, but I'm willing to give it a try and see what happens," he shrugged.

And that was it. We had agreed to embark on a journey to live life without modern technology. That meant unplugging our iPods, laptops, televisions, and digital cameras and disconnecting from e-mail, cell phones and the Internet for personal use. Not an easy tasks for most Americans, especially difficult for us -- a Silicon Valley family dependent on our beloved technology.

The Questions

We entered into this experiment not knowing what the outcome would be, but there were some questions that we were hoping to answer by the end:

1. How will doing away with technology affect our time, money, relationships and the environment?

2. Will we gain quality family time by turning the TV off and instead play games, go on walks, or do art projects?
3. How will life without technology affect our five-year-old daughter who has never known any other way? Will she protest?
4. Does technology save time or waste time?
5. Is technology an expense we can live without or does it help save money?
6. Does technology hinder relationships by limiting face-to-face communication? Or does it improve relationships by making it possible to reach friends easily and instantly?
7. Will a life without technology bring more intimacy to a marriage? Or will it cause distance by not being able to keep in touch constantly?
8. Will living a life without e-mail, social networking, and Internet access cause a strain on friendships due to going against the mainstream? Will we become social outcasts? How will we handle the peer pressure?
9. Does technology help the environment by using less paper? Or does it hurt the environment by wasting electricity?
10. How will we handle the inconveniences of not having the Internet to look up phone numbers, movie listings, get directions or order items online?

Who We Are

Living in Silicon Valley for the past twelve years, technology has drastically influenced who we have become.

The Mom (Sharael):

- A former Web producer who worked in the high-technology industry at its height in the late '90s and has built many websites for high-technology companies such as CNET.com, ZDNet.com and ZDTV.com.
- A professional digital photographer who utilized technology in portrait, fine art and commercial photography.
- An online journalist focusing on technology and the outdoors, with articles published on numerous websites and in various national print publications.
- A mom -- using technology to buy/sell children's items on Craigslist or eBay, asking for support/offering advice through Yahoo Groups, attending Mom's Club events through Evites, receiving parenting newsletters through e-mail, searching parenting Websites for information. Guilty of using TV as a babysitter.

- Taken several community college classes online, including Web production classes.
- Laptops always open. Internet and e-mail always on. iPhone always within reach.

The Dad (Jeff):

- Marketing executive in the medical device industry.
- Facebook addict.
- iPod/iTunes/iPhone fanatic.
- Grossly large iTunes library.
- Nightly TV/movie buff.
- Laptop always open in the evenings.
- Loves to find a bargain (especially on mountain bike parts) on eBay/Craigslist.
- Can't imagine life without Pandora.
- Calls family on his cell phone (hands-free, of course) nightly during his commute home.

The Child (Katelyn):

- Almost six years old.
- Has grown up with technology. Knows no other way.

- Has her own iMac with software games. Has used keyboard and mouse since age two.
- Has her own purple iPod with her own music: *Hannah Montana*, *Camp Rock*, and *Andy Z*, mostly downloaded from iTunes.
- Watches movies on mom's iPod or laptop while traveling.
- Has her own digital camera – a *Hannah Montana* one, of course.
- No Wii (thank goodness), but loves playing it when visiting relatives.
- Movie buff. Owns way too many DVDs. Requests to watch a movie after dinner most nights. Enjoys modern movies, as well as classics.
- Watches TV daily for at least an hour, usually something On Demand.
- Enjoys playing games on children's websites.
- Plays games on mom's iPhone.

What We are Unplugging:

- 1 digital video camera
- 2 iPhones, 1 Blackberry, 1 cell phone
- 2 TVs

- 4 iPods
- 4 digital cameras
- 5 computers
- 6 social networking accounts
- 7 domain names
- 7 blogs
- 8 e-mail accounts
- 9 online photo storage accounts
- 55 DVDs
- No CDs -- all digital music, except Katelyn's books on CD
- Extensive iTunes library with music played wirelessly from our computer in the office to Bose speakers in the living room
- All banking done online

The Rules

Since "technology" is such a broad term, we felt the need for some guidelines for our technology sabbatical. Here's what we came up with:

1. No TVs in our house, but if it's on somewhere else, we can watch it, except at the gym (since we're there nearly every day).

2. Keep a cell phone handy for emergencies only. Don't answer unless it's Katelyn's school, a babysitter, or Jeff/Sharael (who knows not to call unless it is an emergency). Jeff can use his work cell phone for business calls only.
3. Can use only a 35 mm film camera. No digital cameras, including video.
4. No video games allowed, unless at a friend's house.
5. No iTunes or iPods. Radio or CDs (we don't currently own any) only.
6. No computer: This is the BIG one! This means no:

- E-mail
- Internet
- Buying/Selling on Craigslist or eBay
- Banking online
- Social networking: Facebook, MySpace, Twitter
- No maps with directions
- No word processing application (a typewriter instead)

7. Technology can be used when it pertains to work (since we still need to make a living), but not for personal use and not at home. E-mail/Internet/Word processing can be used on a

computer at the library or workplace, if needed for work purposes.

The Implementation

We chose national Turn Off TV Week, April 20-26, 2009, to kick off our unplugged lifestyle. Rather than give up everything at once, we decided to gradually wade into it and ditch one technology at a time as follows:

- Day 1: TV
- Day 2: Cell phone
- Day 3: Digital calendar
- Day 4 eBay/Craigslist
- Day 5: Digital photography/video
- Day 6: Online banking
- Day 7: Gaming
- Day 8: iTunes/iPod
- Day 9: Social networking: Facebook, Twitter
- Day 11: E-mail
- Day 12: Internet
- Day 13: Word processing

A Final Note

We hope this book inspires others to take a look at how technology affects their relationships, time, finances, and the environment...to become aware of their use of technology and to try to cut back in certain areas in an effort to improve their lives. Ideally, *A Year Unplugged* will spark conversations about limiting technology so that it becomes socially acceptable to not have a TV in your home, not always answer your cell phone, and not immediately respond to e-mail or texts, but rather save money, time and the environment while really getting to know loved ones, friends, and ourselves.

Month One: Letting Go

"Television has proved that people will look at anything rather than each other."
--Ann Landers

Day 1: It oddly felt like Christmas morning...in April. Katelyn came rushing out of her bedroom to play with her new toys. Her face glowed with delight when she saw the much coveted play "market stand" sitting in place of the TV...complete with plastic oranges, peas, cookies, turkey, and a pretend cash register, which included a plastic credit card, in case the shopper runs out of the fake cash. Consumerism starts young.

"Oh, Daddy, thank you," beamed Katelyn.

Wait a minute. Who was up until midnight putting together the "easy to assemble," "no tools required" market? *Don't I get any credit?*

The market, and a slew of board games, was purchased the night before, in preparation for dealing with a five-year-old without a TV.

At Toys "R" Us, our basket overflowed with all things girly. Although we tried to emphasize "family" games and activities, somehow we ended up with a pile of Tinker Bell, Barbie and Thumbelina toys. No Lincoln logs, racecars or train sets for Daddy. We did manage to sneak in a Don't Wake Daddy game, Clue Jr. and some Crayola art supplies.

While the cashier rang up our mound of games and toys, I questioned whether we should be spending so much money on things that seemed frivolous. With a grand total of more than $300, I felt the need to justify the purchase to myself, and the cashier, who could obviously tell the gifts were not a surprise since Katelyn was clamoring for each item as it was placed into the shopping bag.

"We're getting rid of our TV," I explained. No response. *Didn't she hear me?* This is a life-changing event for us. At least acknowledge it. Nothing.

***

With Katelyn off to school, I am alone with my thoughts. My mind is racing as I enter a life without TV. No *Rachael Ray*. No *Today Show*. No TV while running on the treadmill at the gym. No TV to keep me company while I work from home. Oh, it sounds so boring. *How will I survive?*

***

Pulling into our driveway, I notice the iris and tulips blossoming. They encircle the base of our tree like a wreath. The air smells sweet. The chill in the air has gone into hibernation. It is spring. A time for growth.

Katelyn hops out of the car like a bunny and races to the house. Seeing Daddy's car in the driveway, she flings open the front door and enthusiastically yells, “Daddy!” when she spots Jeff sitting at the kitchen table, laptop open.

“Hey, my love,” he greets her warmly as she tackles him with love and adoration.

Jeff decided to work from home on our first TV-less day. He pulled himself away from his computer, and his dozens of unanswered e-mails, to enjoy time with his beloved little “mouse.”

Afternoon snack: big, red, juicy watermelon. Katelyn joined her daddy on the front porch for this soon-to-be-summer indulgence. With juice dripping down her chin, the sleeve of her Little Mermaid shirt seemed like the best option to prevent it from making its way to her pants. Jeff, watermelon wedge in one hand, Katelyn in the other, was reveling in the moment. What a nice image. Normally, we’d have the TV on after school. She hadn’t even asked about it.

According to Nielsen Co. the number of minutes per week the average child watches television is 1,680. The number of minutes per week parents spend in meaningful conversations with their children is 3.5. No wonder our children can relate to Mickey Mouse better than us at times.

Disney, Nickelodeon and PBS are raising our children. Until we unplugged, our family was true to the statistics. Although I'm not sure what the definition of "meaningful conversation" is. Does that include discussing the nutritional value of green beans (meaningful to me) or does it mean talking about Hannah Montana's new haircut (meaningful to Katelyn)?

With the average youth watching more than 1,500 hours of television per year, think of what our society could become if TV time was replaced with something more productive. Could we reshape our country by turning off the TV? Interestingly, 73 percent of parents reported they would like to limit their children's TV watching. Then why don't they? Because they don't know how to interact with their children? Because stay-at-home moms need a moment of peace to prepare dinner? Because they're tired from a long day of work? Because they hate saying no to their children? Or because it's become a habit that is never questioned. TV is part of life. Why turn it off?

Katelyn's TV-free afternoon with Daddy was short lived. Fifteen minutes after returning home from school she said, "Can I watch something?"

Ten minutes later, "Can I watch something?"

Five minutes later, "Can I watch something, pleeeease?"

Noticing her new toy sitting on the ottoman just begging to be played with, Katelyn uttered, "Oh, I'm just going to play with Thumbelina instead."

But then, ten minutes later, "Can I watch something? I'm bored!"

"Let's do an art project," I said with a sigh under my breath.

I opened a box of instant mashed potatoes, mixed in some water and told Katelyn, "Look, it's cloud fluff, like on *Charlie and Lola*."

She giggled, "No, Mommy, what is it, really?"

Clearly, I needed to incorporate more vegetables into her diet.

"Let's play animal rescue. This is an avalanche and we need to find the dogs buried in it," I said, grabbing a box of plastic dogs from her toy box and shoving them deep beneath the mound of mashies.

"Forget the dogs," Katelyn said. "I just want to play in it."

She was thrilled with the thought of playing with her food -- and not having to eat it. Maybe not having a TV was going to be a good thing. It forced me to spend time doing a project with Katelyn where we both had to use our imagination and creativity.

***

Finally, a moment alone with my husband and no TV. How did we spend it? I folded laundry in the bedroom -- a boring task without *Entertainment Tonight* to keep me company. Too quiet. I spotted my clock. *Does it have a radio?* I never bothered to notice. Nope. Note to self: buy a clock radio for the bedroom.

As for Jeff, he spent his first TV-less night working on his bike in the garage -- *with* a radio. So much for quality time together.

One day down. Only 364 more to go.

Day 2: I awoke with a headache. *Could it be withdrawals?* I feel like an addict giving up drugs. I'm nervous, preoccupied, not thinking clearly. *How will I make it with no TV...and now no cell phone either?*

According to Robert Kubey, Rutgers University psychologist and TV-Free America board member, millions of Americans are so hooked on television that they fit the criteria for substance abuse and dependency, which includes using TV as a sedative, indiscriminate viewing, feeling loss of control while viewing, feeling angry with oneself for watching too much, inability to stop watching, feeling miserable when kept from watching. No wonder we have such a hard time turning it off.

***

Katelyn is sound asleep for the night, snuggling her stuffed baby and no doubt dreaming about princesses. Jeff gets a sudden surge of energy and decides to go workout.

"How late is the gym open?" he asks me, digging through his dresser drawer for some shorts.

"Ten, maybe," I say.

"Do we have a phone book?" he asks.

"Yeah, in the kitchen, near the phone," I tell him.

As he lugs the book onto the kitchen counter and thumbs through the thin pages, he says, "This is the first time I've opened the phone book in years."

Unfortunately, without the Internet, I assume we will become quite accustomed to using the phone book.

The San Jose phone book has one of the largest distributions, nearly one million. Seems like a waste of paper when most people get their information online.

Regardless, advertisers dish out $282 to $1,288 a month to place an add in the AT&T phone book with the "Original Yellow Pages." Ads can be purchased just in the phone book, or also on YellowPages.com, which includes a free five-page website that AT&T will design and host, with a guaranteed 720 hits a year. The ad will also be promoted on Goggle, Yahoo, MSN.com and 411.com. So, it seems, even the phone book has gone high tech.

***

Katelyn is rummaging through her closet filled with sparkly pink clothes, looking for something fashionable to wear to kindergarten.

"Mommy, is it going to be warm today?" she yells to me.

"I think so," I replied from the nearby bathroom, my mouth half full of toothpaste as I lean over the sink to rinse.

"As warm as yesterday?" she continues.

"I don't know," I mumble.

"Check your iPhone," she says.

Boy, this is going to be a long year. Sigh.

***

In preparation for our year unplugged, I decided to stop paying for my monthly iPhone bill since I would only be using it for emergencies. I went to the phone store to see what my options were.

"I'd like to switch my plan from my iPhone to my old cell phone," I told the salesman.

"Is your iPhone not working?" he asked. "Because you can upgrade to a new iPhone."

"No, I just want a phone to use for emergencies only," I replied. "I don't want to pay for all the bells and whistles that I won't be utilizing."

I handed him my old flip phone and he examined it and started tapping away at his computer screen.

While waiting, I overheard a woman my age talking to another sales representative. "Can I get some sort of family plan?" she asked in a huff. "My cell phone bill is over six hundred dollars a month. Oh, and while I'm here, do you have any new smart phones out? I think I'll pick up a new model for my thirteen-year-old son." The price of technology.

Eventually the salesman that was helping me looked up from his computer screen and said, "It turns out that we don't offer service for this type of phone anymore."

It appears that most customers don't hang onto their old cell phones and try to get them reconnected. So, what *does* happen to all those outdated phones? Some get donated to people with disabilities, to military soldiers or to shelters for battered women. Some get recycled through the world's largest cell phone recycler, ReCellular, who collects 25,000 cell phones daily. And, unfortunately, some end up in our landfills -- leaking toxic materials, such as lead, mercury, arsenic, cadmium, chlorine and bromine into our soil and water supply. Throwing a cell phone in the trash is illegal in California, New York and Main.

"What's the cheapest phone I can downgrade to?" I ask the salesman.

"Uh…we don't do downgrades," he said looking perplexed. "You can terminate your contract for $175 and purchase a new (non-smart) phone for $35 that will have a lower monthly fee."

"So, I have to pay almost $200 for a phone that is *just* for making phone calls…no texting, no Internet?" I said painfully. Much to my dismay, it was the only option.

In 1989, a Motorola MicroTAC flip phone went on sale for nearly $3,000. These days, the only person paying that much for a mobile phone is President Obama with his spy-proof Blackberry. I suppose $200 was a bargain.

Upon driving home, I kept thinking I heard my phone ring. The silence was eerie. Some people claim to suffer from Phantom Vibration

Syndrome in which they feel or hear their phone vibrating, even when it's not with them.

I unwrapped my new phone and created a voicemail message that said, "I'm not checking this voicemail. Call me at home." We'll see if it works.

Day 3: "Mommy, I know what we can do for Earth Day," Katelyn said proudly. "Pick up trash, recycle, save a tree so that the world will not be cut down. And when you brush your teeth, seal the *taplet* tight so you save water. And don't clog the toilet. The Earth needs *some* toilet paper, but not too much. I saw that on a commercial."

Normally, for any holiday, I'd search Comcast On Demand for TV specials. Certainly Dora would be planting a tree or Mickey Mouse would be recycling at his Club House. Not this year. I do feel good not wasting electricity on TV.

***

Katelyn came bouncing out of her classroom, her blonde pigtails flapping up and down like a puppy-dog's ears.

"Guess what?" I say, bending down to her level. "Since it's Earth Day, want to see the *Earth* movie at the theater? It's like our *Planet Earth* DVDs."

"Nah," she said. "I don't care for those movies."

Damn. Now what are we going to do for an entire afternoon without a TV?

"Oh, come on. It will be fun," I pleaded. "We can get popcorn and candy." Bribery is always an option.

"Well, OK, but only if we can go to the market to pick out my favorite candy...toffee (taffy, actually, but I haven't bothered correcting her)."

"Sure," I smiled victoriously.

At the theater, we stepped up to the counter to buy our tickets. I read off the list of movies playing: "*Earth* 3:30, *Sunshine Cleaning* 3:45, *Hannah Montana* 3:40"...0h, *shit, did I just say that...out loud? Here it comes...*

"*Hannah Montana*!" Katelyn screamed, jumping for joy. "I want to see that!"

The *Earth* movie didn't stand a chance.

"Whatever," I reluctantly, but willingly agreed, digging through my purse for my wallet.

I'm not sure if I'm overly emotional with having given up TV or what, but I actually cried during the *Hannah Montana* movie. Yes, she is overly commercialized, but the bottom line is that Miley Cyrus was a little girl with a dream of becoming a singer and now she's living that dream. I want the same for Katelyn, minus the flashy clothes.

It has been reported that Miley Cyrus (aka *Hannah Montana*, for those of you who have not been blessed with a daughter under the age of 12) will most likely be a billionaire by the time she reaches age 18. She seems to be on track with her *Hannah Montana: The Movie* film bringing in nearly $80 million in box-office receipts.

Disney Chief Executive Robert Iger said Disney's mission is, "To create high-quality content and apply innovative technology to raise the level of consumer experience in a way that differentiates Disney."

Considering the fact that you can go see a *Hannah Montana* movie wearing your *Hannah Montana* T-shirt, pants, underwear, socks and shoes, while munching on your *Hannah Montana* fruit snacks and sipping water from your *Hannah Montana* water bottle, and applying your *Hannah Montana* lip gloss, I agree that this definitely raises the level on “consumer experience.” Don't forget to pack your *Hannah Montana* toothbrush and toothpaste.

***

Today my sleek, digital, all-encompassing iPhone made way for a new bulky Franklin and Covey “Organizer 365.” Back to keeping track of appointments and addresses with pen and paper. Not something I welcomed due to the fact that my new organizer required me to purchase a new, larger purse. I am not a big purse kind of gal.

Inside my organizer was a message that read, "*In this high-technology world, information has become the most valuable commodity. To make the most of information, you need to put it to work. That means being able to carry it with you and refer to it as your day goes by. Your organizer makes it possible for you to do just that.*" But it doesn't have GPS!

In addition to my organizer, I purchased a wall calendar...not an easy task in April. Hopping from store to store, I felt like I was on an Easter egg hunt.

Day 4: "Katelyn, want to go to the park?" I asked on a sunny afternoon, since I could no longer spend my time selling and buying on Craigslist.

"No, I just want to go home," she replied, an answer that is ingrained in her like dye soaked into the deep crevices of a fine fabric.

Since the Disney Channel got its clutches on Katelyn at the impressionable age of two, this has been her standard response when asked to do anything other than go home after school.

Even though the American Academy of Pediatrics recommends that children under the age of two not watch television at all, we were unknowingly influenced when we received a *Baby Einstein* video at my baby shower. With Katelyn still in the womb, we already had her television habits established. After all, with a name like Einstein it *must* mean it would make our baby would be smarter.

Baby Einstein, owned by Disney (of course), challenged the AAP by stating, “While we respect the American Academy of Pediatrics, we do not believe that their recommendation of no television for children under the age of two reflects the reality of today’s parents, families, and households.”

However, the *San Francisco Chronicle* recently reported that Baby Einstein would be issuing refunds for their videos to avoid a class action lawsuit aimed at “unfair and deceptive practices.” So, I guess that means watching those videos won’t get Katelyn into the GATE program after all.

Further trying to convince Katelyn to get outdoors, I say, “How about if we stop and get an ice cream on the way to the park?” (again with the bribery).

“Well, maybe if we get a gelato instead,” she said with raised eyebrows, like a fisherman seeing if I would take the bait.

“You win,” I said, when in fact I felt like the victor.

Once at the park, Katelyn complained, “It’s too cold (only 70 degrees), this new park isn’t any fun. I don’t feel like playing.”

I knew what I had to do. I walked across the tanbark and plunked myself down onto the swing and yelled to Katelyn, who was perched awkwardly on a bench that was so high her little feet dangled in mid-air, "Look at how high I'm going!"

Watching curiously, she finally replied, jumping down from the bench, "I want to try!" And, of course, added, "I bet I can go even higher."

The AAP also recommends active play to develop mental, physical and social skills.

When I was a child, growing up in Eureka, California in the '70s, the only TV I remember watching was Saturday morning cartoons: *The Flintstones, Road Runner*, and *Bugs Bunny*. Other than that, I played outdoors with my friends.

Our playground was a field with grass taller than we were, trees begging to be climbed, caterpillars to collect, a Redwood forest with perfectly hollowed out tree stumps for hide and seek, and beaches for building sand castles or collecting shells. Although I feel we have done a good job at promoting a love for the outdoors to Katelyn, with annual camping trips, family hikes and beach days, she is still more comfortable indoors.

In his book, *Last Child in the Woods*, author Richard Louv states, "One might argue that the Internet has replaced the woods, in terms of inventive space, but no electronic environment stimulates all the senses. So far, Microsoft sells no match for nature's code."

Now, if only we could convince our kids that playing tennis outdoors could actually be *more* fun than playing it on the Wii.

Day 5: As the smell of homemade risotto filled our kitchen, Katelyn sauntered over to me while I stirred my masterpiece.

The nightly request, "Mommy, I'm bored. Can I watch something, please?"

"Not now, it's almost time for dinner (*and we have no TV, in case you hadn't noticed*)," I said.

I just about dropped my wooden spoon when she amazingly and quietly walked off and found something to do on her own -- draw on her art easel -- without mommy having to bribe, beg, poke, prod, threaten, or plead. I don't know that I've ever seen her behave that way. Could it be that not having a TV was sparking her creativity and encouraging her independence? Time would tell. If I had the option, I would have grabbed my video camera to tape her being so creatively independent, but the video camera and our digital cameras have been shelved.

Day 6: Katelyn and I are on our own...for two days. No Daddy, no TV. Jeff's away on a short business trip, which was inevitable. The plan...stay busy. First, a "craft club" play date with the Mom's Club...basically a time for moms to chat about the ups and downs of motherhood while their young ones glue sequins to construction paper and snack on animal crackers and applesauce. As chaotic as these dates can be -- imagine twenty toddlers amped up on sugar wielding fully soaked paintbrushes -- I live for these moments.

Although Jeff is a very hands-on dad (lucky me!), even he doesn't fully understand what it means to be in the trenches of motherhood. It's refreshing to be among moms -- who know what it's like to live in sweats for three years because "why wear anything nicer since it'll just be covered in spit-up, mashed bananas, or paint by the end of the day?"

Only moms understand the uneasiness (at least for me) of pulling out your boob in a public place because "that's what God created them for." It takes a mom to truly appreciate that buying yet another toy for my child (who has too many already) is not a luxury item, it's a survival tool. If that additional toy can allow me one second to take a sip of my morning tea before kicking into "entertainer of the year" for ten hours straight until Daddy gets home to relieve me, then I have no problem spending the money. Although, keeping track of my finances would become more daunting without access to online banking. I cherish my time with my mom friends.

Next stop on our "night without daddy" to-do list...dinner at Katelyn's favorite restaurant, that we save for special occasions...McDonald's. Or, as we refer to it, Ronald Old MacDonald's. There's a reason it's called a

"Happy Meal." It's not because it makes the kids happy. It's because it makes the moms happy that they don't have to cook! But there's a trick to it. I've learned how to buck the fast-food system and create a "healthy" Happy Meal: Cheeseburger (hold the meat and sauce), apples (no caramel dip) and plain milk. Katelyn is none the wiser. French fries and soda have never touched her perfect princess lips.

But this time, there was something different about our trip to Mickey D's. Something unnoticed before. A TV! *Oh, joy.* I felt guilty for indulging, but couldn't help myself. I squinted at the closed caption, trying not to miss a word, as Katelyn fumbled to open her Happy Meal treasure. Maybe we'd have to eat at McDonald's more often?

***

Keeping the house tidy was difficult without TV to entertain Katelyn. One day while dusting, Katelyn was watching me like a hawk, curious what I was up to.

"You know, Mommy, I saw a commercial for a duster-thingy that does a much better job than yours," she said.

I thought ninety percent of the time Katelyn watched shows from the DVR with the commercials fast-forwarded. It puzzled and bothered me to hear her references to commercials. Maybe her TV viewing wasn't as commercial-free as I thought.

According to a report by Strasburger in 2001, the average American child may view as may as 40,00 television commercials every year. The Campaign For a Commercial-Free Childhood is trying to change that by "supporting parents' efforts to raise healthy families by limiting commercial access to children and ending the exploitive practice of child-targeted marketing" through public awareness, advocacy, research, and collaboration. Take a look at your shopping list. How many items did your child suggest because of an ad on TV?

CCFC claims that advertising and marketing exacerbate childhood obesity, eating disorders, youth violence, sexualization, family stress, underage alcohol and tobacco use, rampant materialism, and the erosion of creative play. Suddenly, the latest commercial for the Barbie Dream House seems so sinister.

Day 7: An entire week without TV. Katelyn has inquired about it, but hasn't thrown a fit. Planning has helped...keeping busy, staying out of the house, being creative, trying new things. Not having a TV has helped us break out of our routine and expose Katelyn to new activities. I'm just grateful we don't own a Wii or Nintendo DS...depriving her of TV was difficult enough.

***

We need more toys. How could I possibly think this? But without a TV, I feel nervous that we won't have enough things to keep Katelyn entertained. She gets bored with new toys so quickly. I need to have some new backups on hand to stop the potential whining about not having a TV. She already has more toys than she could possibly play with, typical of Silicon Valley, and an only child. Yet I continue to buy more. It's a weakness.

In her book, *The Price of Privilege*, author Madeline Levine discusses the affect that materialism has on children. "Materialism is not only about having shallow values: it is also about how easy it can be to choose the simple seduction of objects over the complex substance of relationships. Materialism sucks the life out of purpose and altruism as kids become increasingly self-centered to the needs of others."

I guess buying yet another Barbie for Katelyn means I am being shallow and putting her at risk to become selfish. And I thought I was just buying a toy.

***

Enjoying a cup of morning tea at my favorite hangout, the Los Gatos Roasting Company, nearly every table is filled with casually dressed customers tapping away on their laptops while sipping a steaming latte. Chatter fills the air and I can't help but eavesdrop on conversations that inevitably address the technology-enraptured area that we live in. When the loud coffee grinder was turned on, a man sitting nearby said to his friend across the table, "I can't hear you. Can you just send me a text?"

At one point a mom with toddler twins was trying to have a conversation with her friend. Her children were whining and crying loudly. Her solution...her iPhone, of course. A *Dora* video and the kids are quiet and happy.

Dora was always welcome in our home. It was my intent to teach Katelyn Spanish, but without the time or energy to do so, I turned that responsibility over to Dora. And it worked. Katelyn can count to ten in Spanish and knows what *vamanos* means, "Let's go."

According to *Dora the Explorer* creator Chris Gifford, the show is based on Howard Gardner's ideas about multiple intelligences. Every episode incorporates seven different learning intelligences based on logical/mathematical, musical/auditory, and bodily/kinesthetic.

The show is scripted so that kids use their intelligences to help Dora and Boots. So apparently Dora isn't all fun and games after all, and children can actually learn something from TV.

In his book *Unplug Your Kids*, author David Dutwin states, "With the proliferation of 24-hour children's programming via Nickelodeon, PBS, Sprout, Noggin and Disney Channels, and other stations, it is of substantial concern whether all programming is created equally. It is not."

Dutwin further explains that the following children's programs have been proven educational: *Sesame Street, Mr. Rogers' Neighborhood* and *Blue's Clues*.

To Katelyn, Elmo may as well have been a rock star. She was a huge and devoted fan. When she was two, we bought her first pet, a goldfish. She named it Dorothy, because that was Elmo's goldfish's name.

The day Katelyn announced, at the tender age of three, that "*Sesame Street* is for babies," was a sad day in our home. *How could she be growing up so quickly?* I wasn't ready to say goodbye to Elmo and his friends, who had become like family to us.

Author of *The Tipping Point*, Malcom Gladwell said, "Virtually every time (*Sesame Street*'s) educational value has been tested -- and *Sesame Street* has been subject to more academic scrutiny than any television show in history -- it has been proven to increase the reading and learning skills of its viewers. The creators of *Sesame Street* accomplished something extraordinary, and the story of how they did that is a marvelous illustration of the second of the rules of *The Tipping Point*, the Stickiness Factor. They discovered that by making small but critical adjustments in how they presented ideas to preschoolers, they could overcome television's weakness as a teaching tool and make what they had to say memorable. *Sesame Street* succeeded because it learned how to make television sticky." No wonder children are glued to the TV.

***

Time to move Katelyn's iMac to the garage. As a child growing up in Silicon Valley, it just doesn't seem right to take her computer away. Will she be chastised in school for not being proficient with the mouse?

Katelyn's first experience with the computer was at age two. As I ogled over the latest laptops at the Apple store, the large screens with Elmo, Dora and Cat in the Hat, vying for her attention, instantly mesmerized her.

She tugged at me and said, "Was dat?" I led her over and perched her on a soft round stool. As she teetered on the seat, she began pecking at the buttons like a bird, her tiny fingers barely strong enough to press the keys.

She sat in wonder as the characters on the screen begged her to solve their problems. I grasped her hand and placed it on the mouse and tried to explain that moving the mouse would also move the cursor on the screen. One trip to the Apple store and her vocabulary tripled...keyboard, mouse, cursor, return button, and software. She was hooked.

Day 8: Katelyn screamed uncontrollably. “Mommy, don’t go. I’ll miss you,” she said sobbing and doing the death grip on my leg as I scrambled through my purse in search of my car keys.

My body was wrought with guilt, down to my core. I was perplexed as to why Katelyn was having such a severe reaction to me going out with my friends for the evening, something I’ve done many times before. Her eyes wet with tears, Jeff picked her up and held her tightly.

“What should I do?” I asked Jeff desperately, feeling torn between having some mommy time and staying with my daughter who so desperately needed me.

“Just go,” Jeff answered reassuringly. “We’ll be fine.”

“Katelyn, I’ll be back soon. I’m just going to the movies,” I told her.

“But I need you,” she cried.

“I love you,” I said kissing her head.

As I drove off, my heart ached with confusion about whether or not I made the right decision. I had to stop myself several times from turning around. She was with Daddy. She’d be fine.

What was going on? This reaction was so out of character for her. She’s Daddy’s girl. Since when does she care if Mommy is around? The only thing that came to mind was that we had more time together since getting rid of the TV. I was spending most afternoons giving her my undivided attention. TV-time was replaced with tickle-time. We were learning what quality time meant, bonding in a way I had not seen before.

The AAP urges parents to limit children's exposure to television over concern, for one, that television will inhibit parent-child bonding rather than encourage it.

Living without a TV was having a dramatic effect on us. And it had only been a week.

Poor Jeff, not only did he have to deal with Katelyn's meltdown, but he was having a bad day of his own. Today was the day we said goodbye to our iTunes and iPods. With his love of music, this day was not one that he welcomed. He would no longer be able to spend time with Dave Grohl or Dave Matthews whenever he wanted.

***

Day 9: "I'm logging off." That was my last message on Facebook...at least for the next year. There goes my social network, but with 360 million users, somehow I don't think Facebook will miss me. It figures that I'm logging of just when the "Causes" app has been released, allowing Facebook users to "support and organize campaigns, fundraisers, and petitions around the issues that impact you and your community."

Day 10: *Yippee!* The *San Francisco Chronicle* arrived today. Finally, some insight to the outside world. I enjoyed the feel of the paper between my fingers as they became smudged with ink and had forgotten how nice it is to sit and leisurely read the news, rather than getting snippets of it online or watching the evening news at bedtime, which usually gave me nightmares.

Just over a week into our technology-less lifestyle and one unexpected side effect is that I'm sleeping much better. Without a TV, Jeff and I are usually in bed no later than nine, rather than 10 or 11. So, we are getting more sleep.

Taking the TV out of our bedroom has made the room more conducive to sleep. Without being bombarded with noise and fast-paced programming before drifting off to sleep, the quality of my sleep has improved.

Dr. Germaine, an assistant clinical professor of pediatrics at Yale University School of Medicine, reports, "Having a television in the bedroom causes delayed sleep time. It is harder to settle down after watching something interesting. It disrupts the ability to get into deeper sleep and can even change vital signs so that it's not relaxing."

Although these are proven facts, parents still brush them off like sand on their feet. "It's hard to limit children's television viewing with parents' busy schedules," Dr. Germaine added, "It's part of our society, but at a minimum, co-view television programs with your kids. Filter what they are watching."

Lying in bed, Jeff said his usual, "Hand me the remote" -- this time with a snicker. I instinctually reach for it on the nightstand, quickly realizing that it's not there. We laugh. This would take some getting used to.

Day 11: We logged off e-mail tonight. Gulp! No e-mail for an entire year, except for work purposes (according to Rule #7). What if I miss something important? An invitation to go wine tasting, a sale at the Gap, coupons for Lunchables at Safeway, the offer to be a fill-in player for Bunco night? These are serious.

I decided I needed to add emphasis to my away message to ensure that people call me instead, "If you're reading this reply, it means I'm not going to get the e-mail you just sent me."

Yesterday, I e-mailed my manager to request a leave of absence as an online journalist (but will still write occasionally for print publications). After all, how could I possibly feel what it's like to live a no-technology lifestyle while being required to upload articles weekly online? Unfortunately, I did not hear back from my manager before I unplugged. Will I get fired for not turning in my articles?

Day 12: When Katelyn started kindergarten last fall, it was time to get my pre-baby body back. I managed to lose 15 pounds and get back to weighing 115. My secret weapon...Rachael Ray. Not for her cooking tips, but because on the treadmill at the gym every morning I was entertained by *The Rachael Ray Show.*

Since unplugging, I've been avoiding the gym. Running on the treadmill is so boring without TV or an iPod. However, today I was not in the mood to tackle a run outside in the crisp air, so off to the gym I went.

My gym has at least 14 pieces of cardio equipment with a TV attached: treadmills, ellipticals, and stationary bikes. If only I had access to my Facebook or email...I'd be burning hundreds of calories a day, since once I log on, it's hard to log off.

While working as a broadcast reporter for ZDNet News in 1997, I covered a company called Netpulse. They made stationery bikes with Internet access. You could check your e-mail while working up a sweat. A brilliant idea. But in 1999, after merging with eZone, the company went bankrupt. The cost of the $4,000 bikes didn't allow enough room for profit, plus not many gyms had T1 lines. A company ahead of its time.

Walking past the row of treadmills, I decided to opt for the elliptical. After flipping through the free rack of magazines, I selected *Coastal Living, Wired*, and *Glamour* and attempted to balance on the elliptical while reading. The large-screen TVs that hung on the wall in front of me were too tempting. With no sound, I still kept peeking at them to see if I could get a glimpse of what was being discussed. A flu epidemic...the swine flu. *What if there was an emergency and I had to TV to tune into?*

The memory of the 9/11 tragedies was still vivid in my mind. It was 6 a.m. when the phone aroused me from my sleep on that tragic day. With a sense of urgency, I answered it. Jeff's mom was on the other end, "Turn on the TV. A plane has crashed into the Twin Towers in New York."

I flipped on the news and rushed to the bathroom to get Jeff out of the shower. With a towel wrapped around his waist and still dripping wet, he joined me as we sat astonish at what was unfolding. Sitting on the edge of the bed, we watched in horror as we saw the second tower collapse in a cloud of dust. Tears welled up in my eyes.

"Oh my God, oh my God," I muttered. "All those people." In a matter of seconds, hundreds of lives were gone. The impact of that moment changed the world. Families were broken, businesses crumbled. Peter Jennings kept us informed during the tragedy and dismay that ensued. To have been without a television that day would have meant missing a major piece of world history, witnessing a change in humanity and a widening gap in race relations. *Are we being irresponsible by completely removing the TV from our home?*

Day 14: Our first full day with no technology and, of course, it's raining. Rather than being stuck indoors all day, we headed to Fresno to visit Grandma and Grandpa. Sounded like a good idea, until an hour into the three-hour drive, Katelyn was bored.

Without an iPod to listen to *Hannah Montana*, Mommy's iPhone to play games on or Mommy's laptop to watch *High School Musical* on, she couldn't fathom the torture of two more hours in the car -- even with a backpack stuffed to the brim with coloring books, card games, Polly Pockets, *Barbie* magazines and sticker books.

When I was a kid, we would make the long five-hour drive from Eureka to Sacramento to visit my grandma. Our entertainment was drawing crayons and paper or playing with Barbie dolls as we swayed to *Simon and Garfunkel* on our 8-track while cruising down Highway 101. Inevitably my sister Maggie, five years my senior, and I would fight for space in the back seat. On our journey, we would count red cars, look for license plates from each state, search for every letter of the alphabet in street signs and play car bingo.

To Katelyn, hours in the car normally meant being glued to the computer screen -- engulfed in the latest G-rated movie. As we drove through the countryside, cows dotted the hillsides like sprinkles on a cupcake.

"Let's see how many cows we can count on the way to Grandma and Grandpa's house," I suggest to Katelyn.

She's game. After about 15 minutes -- at 75 cows, she had enough.

Jeff later confided to me, "I didn't even know she could count to 75."

"I'm bored," Katelyn whines as she fidgets in her car seat.

"Just close your eyes and take a nap and I'll wake you when we get there," Jeff said.

Katelyn protests, "No, I changed my mind. I don't want to go to Grandma and Grandpas. It's too far."

Jeff reiterates, "Just close your eyes, my love."

At this point, *I* closed *my* eyes and drifted to sleep.

Upon awakening, I looked over at Katelyn and the *swish, swish, swish* of the windshield wipers had lulled her to sleep.

While visiting my parents, I asked my dad, "Do you still have the old typewriter that I used in Junior College?"

He dug through his garage and emerged with a typewriter covered in a layer of thick dust.

I asked if I could take it home and said, "We have decided to take a break from our computers for a while."

His response, "Good luck with that one."

***

Katelyn held out for two weeks before unleashing her protest of a TV-less life. After entertaining Katelyn with bouncy ball, tickle time, and giving in to playing Clue Jr. yet again, by 5 p.m. I needed to jazz things up. *Hmm. Maybe just a sip of that new organic wine I bought. Well, maybe a half-glass. Maybe a whole glass.*

Jeff read my mind and showed up early at 5:30 p.m. I needed some backup. With dinner done at 6:30 p.m., Katelyn let out an ear piercing, "I really want to watch something!"

The tension was rising inside of me like the mercury in a thermometer. *How did we stave her off for this long? Here we go.*

Jeff tried to console her, "Why don't we play the Don't Wake Daddy game? That's fun."

With a pissy tone, Katelyn replied, "I don't like that game. It's stupid."

It must have been the wine talking, but I heard myself saying, "We could play Clue again."

"No!" she snapped, arms folded across her chest.

"I'm craving apple pie," said Jeff, trying to change the subject.

"I know a great pie shop," I said.

"I don't like pie! I want gelato," Katelyn said.

"We have to take turns, Katelyn. It's not always about what you want to do," I informed her.

"But I want gelatooooo!" she screamed.

OK, when did my child turn into Veruca Salt from Willy Wonka and the Chocolate Factory? Deep breath. Regroup.

"I know the perfect place," I announced. "They have pies, cookies, cake and ice cream."

With a pouty lip, Katelyn said, "Well, OK. We can go look, but if they don't have something I like, then we are going to get gelato instead." She stomped her foot for emphasis. She meant business.

At the pie shop, Katelyn fancied an M & M cookie. Problem solved. We lingered in our faux leather booth, dreading our return to a TV-less home.

***

No TV and still no true quality time with my husband after two weeks. In bed early every night catching up on years of lost sleep attributed to the "parents of a child under the age of three"' phase of our lives.

Normally, I'd get sucked into some reality show that would keep me up way past my bedtime.

According to Nielsen Co., the average American watches more than four hours of TV a day, which means that annually Americans watch 250 billion hours of TV and 49 percent say they watch too much. I was enjoying falling asleep in my husband's arms instead of staying up late to catch up on chick shows.

Day 15: With a typewriter, I am in need of typewriter ribbon. After several trips back and forth to Office Max and Office Depot, I felt like I was on a scavenger hunt. Finding a new typewriter to purchase, surprisingly easy. Finding a clerk that knows anything about typewriter ribbons, not so easy. Upon locating what I needed, for sixty-five dollars, I cleared out the stock of typewriter and correction ribbon.

***

Our new technology-free zone felt good. My anxiety of doing without was replaced with a sense of peace just by decluttering. With our technology gadgets spread throughout the house, they hadn't seemed like much. But filling an entire five-shelf rack in the garage, I realized the mountain of technology had been draining our time, money and energy.

In her book, *Throw Fifty Things Out: Clear the Clutter, Find Your Life*, author Gail Blanke states, "...living in the Information Age doesn't help. We're constantly bombarded from every direction by flying debris in another form: the news, the media, on television, on the radio, on our cell phones, online, and in the air, we're deluged with what too often turns out to be life marble -- *garbage* might be a better word."

Cleaning our home of our technology belongings made me want to purge in other areas as well, masked under the term "spring cleaning." Time to throw out old clothes, old toys, and anything taking up space but not being used.

Day 16: The day I've been dreading is finally here. Katelyn is home sick from school, with no TV. A sick child + no television = torture.

Luckily she slept most of the day. The rest of the day was spent with me reading her endless amounts of books and gladly giving in to her request to eat nothing but pancakes doused in sticky syrup because "I might throw up if I eat anything else."

Day 17: Katelyn is still sick. Good news -- she doesn't have a fever, so she's feeling better. Bad news -- she doesn't have a fever so she won't be sleeping all day. I don't know which is worse, the whining for "Can I watch TV, please?" or the "Mommy, just read this book one more time."

She relentlessly forbids any books on CD. Another book marathon day. How many times can mommy read *Cinderella*? And I used to think her being home sick with a TV was difficult. While reading her *Amelia Bedelia* book for the tenth time, literally, I began laughing deliriously.

"What's so funny, Mommy?" Katelyn asked.

"I just can't take it anymore. Can we *please* just pick a different book? You have three shelves of books to choose from," now I was whining.

"But I like this one, Mommy," she said innocently.

***

It's official. No TV equals better sex. After two weeks of recuperating from a life of never-ending to-do lists, we finally had the energy to spend time being frisky. And we're not just talking obligatory marital missionary position -- we're talking foreplay! A novel idea since becoming parents.

In bed at our new norm, 9 p.m., Jeff and I, for once in a long time, weren't too tired to truly enjoy each other -- organically, not rushed.

With a TV in the house, we'd usually stay up later than intended and crawl into bed too exhausted from a busy day of work and parenting to really focus on each other. Not anymore.

***

Already I'm starting to doubt the strength of my friendships. It's been two weeks and no one has called me. Even though my cell phone is for emergencies only -- I can't help but keep checking it constantly to see if anyone has called. No. *Am I already growing distant from my friends because I'm not as easily accessible? Or is it that we never really communicated by phone anyway and I just didn't realize how dependent I've been on e-mail? Come to think of it -- when's the last time I really sat and chatted on the phone with a friend?* I need to put a plan in action to start calling my friends more often.

Day 18: I've gone from being a professional photographer to carrying a disposable camera in my purse. Who knew there were so many different kinds: compact, zoom, flash, underwater.

While at Katelyn's gymnastics class, I whipped out my disposable camera to capture her in action. Surrounded by a room full of moms, I snapped a shot and winced at the loud *zip, zip, zip* noise the camera made as I advanced the film. *Did think I was too cheap to buy a digital camera?*

I recently saw a young guy wearing a T-shirt that read, "Film is Not Dead," but the Associated Press announced this week that, after 74 years, the Eastman Kodak Co. is doing away with its popular Kodachrome color film because of "failing customer demand in a digital age."

It's amazing how quickly technology can change the world. When I started my professional portrait photography business in 2003, I was using film. By 2006, I went digital. It was much faster and cheaper than shooting film. According to PMA Market Research, digital cameras are expected to gross $5.3 billion in 2009.

While studying photojournalism is college, I was glad that my teacher showed us how to shoot and develop film, but print in the dark room and manipulate photos in Photoshop -- a great mix of old and new technology. Students these days will most likely never see a dark room.

***

Jeff's mom called today, concerned that we didn't reply to an e-mail she sent announcing she had submitted her retirement notice...after thirty years of employment. Not a trivial e-mail.

"It worried me that I couldn't get through to you," she said with motherly concern. "I put a call into your cell phone too, but you didn't get that either, I guess."

Choking up, she continued, "After I submitted my resignation, I just sat in the parking lot and cried. Then I went home and watered my flowers in my bare feet. I'm just glad to know that you're all right. I was really worried when I didn't hear back from you. That's not like you."

We had chosen not to make an announcement to our friends and family about our decision to unplug, since it was really about focusing on our time together as an immediate family.

"What led to your decision to not use e-mail?" Jeff's mom inquired.

Upon hearing the distress in his mom's voice, Jeff might as well have been a dragon with a knight plunging a dagger into his heart.

"We're just trying to get back to basics, Mother," Jeff explained, as I sat close by with my ear pressed firmly against the other side of the phone anxiously awaiting her reaction. "We're taking a break from technology to reconnect with our family. You might not reach us by e-mail or cell phone, but you can always call us at home. I'm so sorry we worried you."

She, forgivingly replied, "I like that idea."

Day 19: *Riiing. Riiing.* I look at the caller ID. It's my mother-in-law. I pick up joyfully, "Hello?"

"Katelyn?"

"No, it's Share," I respond, hating the fact that my voice really does sound like a five-year-old.

"I'm glad I caught you," she said. "I'm having trouble opening an attachment in e-mail. Could you walk me through it?"

I am the go-to person in our extended family when it comes to technology issues. Need to block those annoying pop-ups offering naturally enlarged breasts? Want to use eBay to hock that abs machine you bought to finally get that six-pack, but has been sitting in the garage in an unopened box for six months? Can't wait for the live version of the latest American Idol reject's performance as your ring tone on your iPhone? I'm your gal!

My interest in technology wasn't something I was born with. It was forced upon me by a college professor who, in 1995, required us to have an e-mail account (on Unix, remember that?) to take quizzes sent to us.

The journalism department's computer lab was cold, stark and very uninspiring, like entering a hospital ward (at least that's how I remember it). The air was stale and the wait long. I had better things to do than sit around a computer lab all day waiting to check my e-mail. After all, I was attending the University of Hawaii. My time was better spent learning to longboard, sucking up Mai Tais, or working on my all-over tan.

So, I did what any student with a creative financing flair would do. I took out a student loan and bought my first computer...a Mac. Somehow, I missed the fine print that said I would still be paying if off thirteen years later.

After discovering how to hack into the fraternity's e-mail account and infiltrate their secret society, I was hooked. My love affair with technology blossomed like a tulip in spring.

When Jeff and I started dating in 1997, he would joke with me, "It's a PC world." But once he married me, he didn't stand a chance. Like a missionary trying to convince someone to change his or her beliefs, I quickly converted Jeff to living life as a Mac user.

I took him under my wing, like a frail baby bird just learning to fly, and passed on my computer knowledge and helped him grow into a mature Mac addict, able to soar above and beyond basic computer skills.

Jeff eagerly grabbed the phone from my hand and provided technology support to his mom in Orange County. Afterwards, My mother-in-law again inquired curiously about our no technology lifestyle. Jeff, beer in one hand, spatula in the other, held the phone between his ear and shoulder while flipping burgers on the barbecue.

"I'm still using my laptop for work. It's business as usual, but I won't have Internet access at home," he said. "That means if I have work to get done I'll have to go to Starbucks in the evenings, but the flip side is that I'll spend more time with the family without the lure of my laptop."

My mother-in-law listened intently and replied, "It reminds me of people doing without their TV."

Little did she know that we, in fact, were also doing without our TV. But one thing at a time. We didn't want to overwhelm her with explanations of our decision to unplug from every technology.

"Well, Mother, we're in for a real adventure," Jeff chuckled while checking his burger.

"Next thing you know, you'll be using candlelight," my mother-in-law added lightheartedly. "It'll be quite an experiment."

"I'm not sure if I've bought into it yet," Jeff admitted.

"You don't have to," she responded with a supportive tone of voice. "Just give it a try and see what happens."

As Jeff balanced his monster burger precariously on a too-small bun, he slathered on some ketchup and took a seat at our weathered backyard picnic table. Katelyn and I joined him with our tofu corndogs.

The spring air was warm and welcoming. Bees buzzed by us as if we were a tourist attraction, but thankfully didn't stop for a closer look. I couldn't help wonder what Jeff's mom was thinking about our unplugged lifestyle. Unheard of for technology lovers like us.

"She's really concerned about us," Jeff said with a sigh. "I don't want her to worry."

"Do you think she assumes we are having marital problems or something? That's the conclusion I would jump to," I said, batting away pesky flies.

"Who knows," Jeff said. "It's going to be hard for people to understand why we willingly made such a seemingly monumental sacrifice."

***

As I stirred my pot of homemade vegetable soup, the song *One, Two, Three, Four* by the Plain White Ts played on the radio. Turning up the volume, I started singing to Katelyn and doing hand movements to go along with it. Jeff joined in and grabbed a wooden spoon as his microphone.

Katelyn was giggling hysterically.

"You guys are so silly," she laughed.

After our performance, I turned to Jeff and said, "I wish we could download this song off iTunes." *Pout.*

Day 20: Twenty days and no one has heard my voice from my cell phone...until today. As I was browsing the bookstore for the perfect retirement book to send to my in-laws, it struck me. Somewhere between the *There is Life After Retirement* and *You're Retired...Now What?* books, I felt like I had forgotten something. I frantically dug through my bottomless purse in search of my wallet. Upon opening it, I discovered an empty space where my bankcard should have been. *Oh, crap!* My heart rate quickened as I mentally retraced my steps...drycleaner, Safeway, Target. Yes, this is how moms spend their days, not Banana Republic, movies, lunch at Cheesecake Factory, massage -- contrary to what our husbands might think.

I grabbed my cell phone and dialed 411 to get the number for my last stop, a pizza place. Indeed, they had my card and would hold it in the safe for me to pick up later.

Thinking back on my college days in the mid-'90s, we didn't have cell phones, we had pagers. Back then, our emergencies consisted of getting advice on what to wear to the frat party, needing comfort after having been dumped by the latest loser, or being out of Coors Light.

Upon receiving a page, we'd have to rush to find the nearest pay phone to return the call. Technology has come a long way since the days of pagers and pay phones.

Day 22: "I e-mailed you some photos of the girls," my friend Diedre informed me at Katelyn's school. "You still have your 'away' message on."

"Oh, yeah. I'm taking a break from e-mail," I said timidly, shoving Katelyn's *High School Musical* backpack into her cubby.

Diedre obviously assumed that my "away" message was from our recent vacation. It's called an "away" message because people use it to inform others they'll be away from their office or home -- like on a trip to Disneyland trying to fulfill their daughter's request to "get a photo of EVERY princess," even if it means waiting in lines longer than the Golden Gate bridge while desperately having to pee because you regretfully ordered the Big Gulp-sized Diet Coke at lunch.

Why don't e-mail providers offer the option to set a "technology break" message to let people know that you're around, but taking a break from technology? *Oh, I know why...because who on earth would want to deliberately subject themselves to a life without e-mail?* Maybe once people figure out it's not as scary as it sounds, options will open up.

***

As I approached Katelyn's classroom, her class was lined up outside the door.

"What's going on?" I asked the teacher.

"Since one of the girls didn't want to participate in the Mother's Day performance we did this morning, the girl's mom asked the class to redo it so that she could video tape it because her daughter is now being cooperative," she said.

I grabbed Katelyn's hand to leave.

"Just because one child didn't want to sing, that doesn't mean the entire class should be forced to do it over again," I said loudly enough for the mom to hear.

In the parking lot, I approached her and inquired about her motives.

She informed me, "I told my daughter that not participating is not acceptable and that if she didn't do it the first time, she'd have to do it again later."

"Did you ever think that maybe she didn't want to do it because she was having stage fright?" I asked the mom bluntly.

"No, it's not that. She was just trying to irritate me by not doing what I asked her to do. If I let her get away with it then she would not want to participate the next time either," she said.

Why do we force our children to do things they don't want to do?

I've seen so many parents demanding that their child take photos with Santa at the mall. We tell our children not to talk to strangers, yet expect them to sit on a strange man's lap and tell him their inner most desires.

A child will be screaming hysterically while the parent tries bribing ("I'll give you some candy"), reprimanding ("You better do this or you'll get a time out"), or using humor (making silly faces behind the photographer). *Is the photo really worth tormenting your child?*

On our recent trip to the Grand Canyon, I was guilty of "strongly encouraging" Katelyn to take "just one more photo" at various points of interest. She protested every time and, even now, almost a month later, when asked about the trip she replies, "We had to take too many photos. It was no fun."

Living without the video camera has been the most difficult thing about being unplugged. I'm heartbroken that I will not be able to videotape upcoming events: Katelyn's *Hannah Montana* performance at school, a ballet recital, her sixth birthday, her kindergarten graduation. *Am I letting her down by not having these moments on video for her to look back on when she's an adult?*

***

As I sit at my desk pressing the keys of my typewriter, I enjoy the feel of it and the sound of the keys pressing ink onto the paper. However, because it is so loud, I can no longer work at night after Katelyn (and sometimes Jeff) is asleep. This is the time I feel most creative. After the lights have gone down, the air has cooled, the world has turned in for the night and the only sound is that of crickets serenading me with their lullabies.

Maybe I should move out to the garage. A desk squeezed between the shelf overstuffed with camping gear and the shelf filled with Christmas decorations, old photo albums and books that *I better hang on to because I just might want to read these again someday* (that have been sitting in the garage taking up space for years). The fluorescent lights and lack of heat or air conditioning would be something to contend with.

For now, my cozy, softly lit, spacious, clutter-free office sounds a bit more appealing. My typewriter is doing the job, but I miss spell-check. Instead, I have to rely on a dusty old dictionary with yellowed pages and a musty smell to ensure accurate spelling.

Day 23: Marie's voice came through the phone loud and clear, "You're not strengthening relationships by not being on e-mail. You're actually going to cut off communication. No one has time to make phone calls anymore. It's inconvenient."

Jeff and I listened to his cousin's point-of-view intently over speakerphone. I felt like a child being scolded for not cleaning my room.

"It's not like people can't reach us," I said trying to defend our decision. "Feel free to call us anytime at home."

"The only free time I have during the week is in the evenings and that time is spent with my family," she continued. "The second I walk in the door, they don't want me to jump right on the phone. I check e-mail before the kids get up so that it doesn't interfere with our family time."

She made a good point. Jeff joined in to offer further explanation, "We're just trying to break the habit of spending our entire evenings on e-mail and Facebook. Instead, we've been playing board games together. It's been really nice."

"I commend you for being thoughtful of your family time. I respect that," she continued. "But my feelings are really hurt because it's going to be less contact with you. It's so comforting to get e-mail from Sharael updating us on what you guys are up to. It's like getting a gift in the mail. I will miss that."

"I didn't realize the impact," I apologized.

"I realize that e-mail is impersonal, but by not being easily accessible, you're going to piss people off," she said. "And I think you'll miss out on a lot of stuff. Why not get a new e-mail just for family to use? Otherwise, you should think about relocating from the Bay Area to Orange County because, from the family's angle, what you're doing is failing. I do not like you guys pulling the plug."

Sensing the panic in Marie's voice about not being able to reach us via e-mail, I felt like she was grasping at straws trying to come up with any excuse to try to change our minds.

Jeff eventually ended the conversation with, "You can call us anytime. And we'll do the same." And we hung up.

"Wow, I had no idea that our decision would have such a drastic effect on others," I said to Jeff, curling up next to him in bed. "We are supposed to be doing this to improve relationships, not strain them."

"You should be touched," Jeff said as he reached over to turn out the light. "She obviously cherishes the e-mails you send her about our family."

Drifting off to sleep, I didn't know whether to be sad, glad or mad about my cousin-in-law's reaction. I love her dearly and didn't want to upset her, but I really had hoped that her reaction would've been, "I'm so happy that you're putting your family first and taking care of what's really important and we'll do anything we can to support you."

Marie argued some valid points. I wondered if others felt the same way but just hadn't voiced it.

***

At a recent Mom's Club meeting, the president asked who would like to host the next month's play date -- my hand shot up like a rocket -- an excuse for my annual tie-dye event.

The president said, "You will just need to send out the Evites with the date and time."

Shit. I don't have Internet. What am I going to do? My palms began to sweat at the thought of admitting to the entire group that I'm living unplugged. It's embarrassing.

Weaving through the crowd of chatting moms like a mountain lion stalking its prey, whom could I approach to do my dirty work of sending the Evites?

Not having e-mail or Internet would severely limit my ability to be involved. Something I hadn't thought of.

I can remember when Evite.com launched in 1998. Sitting at my grey-walled cubicle at ZDTV in the SoMa (South of Market) district of San Francisco (also known then as Multimedia Gulch), my co-worker Jeremy swung by to tell me he was quitting to work for a start-up company that would revolutionize party invitations. The thought to jump ship crossed my mind since I have always been a party planner, but I stayed put. The right decision? Jump ahead to 2009...ZDTV is no longer in business. Evite.com, on the other hand, now has over 22 million registered users and over 25,000 invitations are sent out each hour. Now that's what I call revolutionizing invitations.

Day 24: While listening to my only source of entertainment, the radio, I've become close friends with Sandy, the DJ on Mix 106.5 (although we've never met). She is there for me, even when I'm in a bad mood and not much fun to be around. She entertains me by revealing tidbits of relevant, witty information. Today's topic was Twitter.

"I'm so glad that not everyone is obsessed with social networking," she stated. "There comes a point when you just need to detach. Kanye West doesn't like Twitter apparently. He posted a blog saying, 'I'm too busy actually being creative most of the time and if I'm not, and I'm just lying on the beach, I wouldn't tell the world. Everything that Twitter offers, I need less of.'"

Nice to know that I'm not the only person in the world not on Twitter. Another day, Sandy was talking about eBay and *American Idol*, two of my favorite things.

"Have you heard about all the *American Idol* items for bid on eBay? Someone is selling a piece of Adam Lambert's hair and people are actually bidding on it. There are 700 items for Adam."

That's nothing compared to Ian Usher who, in 2008, put up his "entire life" for auction, including his house in Perth, belongings, introduction to his friends and a trial at his job. His life sold for $384,000.

***

"I need to tell you something," I said to the Mom's Club president. "I hope you don't think I'm weird, but we've disconnected our Internet."

"And your TV, right?"

"Yeah, it just got to the point where our evenings were spent with Jeff on Facebook and me on e-mail while Katelyn was watching a movie."

"We're the same way," she replied.

"It's been nice not having it," I admitted.

"Maybe we should do that," she said. "Although, I don't think my husband would go for it."

I spoke up and said, "So, I have a favor to ask. Do you think you could send out the Evites for the craft club for me?"

"Of course, no problem," she said.

Problem solved.

***

Without the patience to workout on a TV-less treadmill, I dug out my book about local hiking trails and enjoyed burning some calories while breathing some fresh air at Rancho San Antonio Open Space.

I have always thought of myself as an outdoors person, but that meant running (fast) with my iPod (loud) on a popular cement running trail near my house (quick and crowded). Living without technology has forced me to slow down and appreciate nature even more.

Making my way up the dusty trail, water bottle in hand, I was enjoying the sights and sounds of nature when I heard a noise*. A growl?* I stopped, looked around and came eye-to-eye with a bobcat perched in a tree. *Should I run?* I slowly backed away, and then ran. *Is he chasing me? No.* I came for a hike to enhance my life, not endanger it.

A few yards down the trail was a woman walking alone with her iPod on. I waved her down and told her, "You might want to take your iPod off. I just saw a bobcat, so just be aware."

"Thanks," she said, wiping the sweat from her brow, "I was just singing and wouldn't have heard it."

***

My friend Maddy came to pick up her daughter after a play date and I suggested that we do a mom's night out. She agreed and said, "Let me look at my calendar."

She then proceeded to reach in her purse for what I thought was an iPhone, but in fact it was an organizer like mine.

When I asked about it, she said, "I just like the feel of it."

I'm beginning to wonder if there are low-technology people out there living a secret life. Jeff has coined the phrase "Minitech" to describe a person who lives with minimal technology. Unlike a Luddite, who is opposed to advances in technology, a Minitech just tries to cut back on their usage of technology. Everything in moderation.

***

Day 25: To avoid any misunderstandings, I updated my e-mail "away message" with something more personal:

"I will not be receiving your e-mail because I am unplugging for a while, in an effort to reclaim quality family time and strengthen relationships through talking in real-time. Sorry for any inconvenience. Call me at home anytime. Sharael".

***

As Macy and I sunned ourselves poolside, she said to me, "I tried to send you an e-mail and I got your 'away message'. What's going on?"

Once again, I would have to explain how technology had taken over our lives and we were taking our lives back. She surprised me by saying, "I grew up without a TV until I was 12. We baked our own bread, made our own Play-Doh. It must have been difficult for my parents to not take the easy way out."

"That's for sure. It's definitely challenging, but we are playing board games in the evenings instead of watching TV now. It's been fun," I said.

Most parents have a difficult time giving up TV because it takes too much thought and planning to come up with alternatives. But others have done that for us. Books, such as *Unplug! 101 Way to Pull Your Kids Away from Television* by Wanda Kanten Hartfield, offer a multitude of activities without a television.

Day 26: In tears, I tell Jeff, "I just can't imagine not videotaping Katelyn's upcoming school performance. She's been rehearsing for months and is even doing a duet."

"Yeah, but when we were kids our parents didn't have video of us," he said, trying to make me feel better.

"My dad did," I replied. "And I cherish those videos...although they've been lost in his garage for over ten years. Who knows if he'll ever find them."

"We'll just take photos," Jeff said, putting his arm around me.

"I'm about ready to pull the plug on this unplugged lifestyle. I just don't know if I can take it emotionally," I weep. "I don't think I can live without videos of Katelyn. It's so important."

Jeff said, "It's up to you, but I don't think you should give up just yet."

I painfully refrained from bringing my video camera. As I looked around the room full of parents, everyone was capturing the moment on video. One mom held a video camera in one hand and a digital camera in the other, capturing the moment on both simultaneously.

As I sat there watching Katelyn sing and dance her little heart out, taking in every detail without the distraction of video taping -- her hip shaking, the deep pink color of her sequined shirt, her golden hair swishing back and forth as she rocked out to the music in front of the *Hannah Montana* backdrop that was messily hand-painted by the kids.

I learned to appreciate the moment and to live (although maybe not happily) without a video camera.

Day 30: I woke up to an empty bed and heavy heart on our nine-year wedding anniversary. With Jeff in Spain and not accessible by cell phone, I am lonely. I miss him, his touch, his voice. I don't want to get out of bed, but alas my alarm clock – Katelyn -- won't let me fritter away the day. By no later than 7 a.m., I am greeted with the usual, "Get up. I'm hungry!"

I drag myself out of my cocoon and morph into a beautiful butterfly -- *ah, what a shower and make-up can do.* My spirits are low. The toll of living unplugged is catching up with me. It's stressful...emotionally, physically, psychologically.

My cousin-in-law has me on her pooh-pooh list, Jeff is not reachable on our anniversary, I miss TV, I hate working out at the gym without being able to “keep up with the Kardashians” while running on the treadmill.

What better way to beat the blues...retail therapy. A trip to Santana Row is in order.

***

My spring-cleaning spree has left a mound of unwanted items in the middle of my garage. Without Craigslist or eBay, we are going to have a good old-fashioned garage sale. Katelyn is ecstatic.

Month Two: My husband Left Me for Double D's

"The amount of control you have over somebody if you can monitor Internet activity is amazing."
--Tim Berners-Lee, Inventor of the World Wide Web

Day 31: Ten minutes to 8 p.m. We just have to see the *American Idol* finale. *Where can we find a TV to watch it?* I called my friend Bree to see what time it was scheduled to start.

"I don't know. I haven't been keeping up with it," she admits. "Let me check the listings on my TV. Oh, it's on at eight o'clock tonight!"

It's too late to call a friend to ask if I can go to their house. And I can't take Katelyn to a bar to watch it. I doubt that it would even be on. *A hotel.*

"Do you think I can find a hotel that rents rooms by the hour?" I ask Bree.

"If you can, it's probably not the type of place you want to take Katelyn," she laughs.

Jeff is out of town on business. I grab the phonebook and start madly dialing hotels nearby for the cheapest rate. While on the phone I'm hurrying Katelyn, "Get your suitcase, pack your PJs, toothbrush, undies, and something to wear to school tomorrow. Quick! We're missing it."

We rush around frantically. I grab my gym bag and stuff it with my PJs, toothbrush, jeans, T-shirt and undies -- forget the makeup.

We speed to the hotel (not setting a good example of being a law-abiding citizen, but hey, it's the AI finale!). On our way, we chant, "AI, here we come!"

Checking into the hotel, the clerk offers us the "recession" price of eighty-nine dollar. Hoping for some pity and maybe an even a deeper discount, I confess that we live just down the street but don't have a TV and just want to watch *American Idol*. He is young and unfazed.

As I open the door to our room, it smells musty, like dirty sweat socks. To mask the stench, I slather my body with my coconut-scented lotion. Katelyn and I snuggle under the scratchy sheets (no 400 thread count Egyptian cotton here). The polyester comforter provides little warmth.

The room reminds me of motels that were within my budget as a starving college student. So much for a fancy flat screen TV. We get a 17-inch wall-mounted behemoth instead. But really it's not about the TV or even about *American Idol*. It's about making a memory. Mommy and Me time with Katelyn. Being adventurous, mixing up our routine -- lumpy pillows and all.

As we watch the contestants sing their hearts out, I reminisce about last year's finale -- well, the pre-finale -- the show-off between David (Cook) and David (Archuleta).

Three days before the show was to air, I was online late at night checking e-mail, when I received one with a subject line "American Idol Tickets." I clicked on it and read that I had been chosen to attend the taping, live in Hollywood. As I re-read it, my heart quickened, but I thought it must be a scam. *American Idol*...live! Better than winning the lottery. I rushed to the bedroom to show Jeff the printed e-mail. His level of enthusiasm paled in comparison. We came to the conclusion that the e-mail was truthful, since I had signed up on the AI Website to win tickets.

I was allowed to bring three guests to L.A. with me. I composed an e-mail and sent it to everyone I knew. How would I decide whom to take with me? I didn't want anyone to feel jolted.

As is the norm in Silicon Valley, at 10 p.m. I received replies instantly. “No” from Ellie -- it's on a work night. “No” from Jennifer -- it’s on a school night. “No” from Elaine -- no money for the airfare. “No” from Lori -- she's not a fan of the show. *Come on people!* Finally, the one and only “yes” -- from Heather. Luckily she was game and her husband was willing to take care of their two kids.

On the road at 5 a.m., I drove to San Mateo to pick up Heather to catch our 7 a.m. flight at SFO. Our marathon day began with pancakes at the airport -- a meal that would unknowingly have to tide us over until dinner.

Folding chairs to sit on while waiting in line -- *check.* Hand-painted signs ("David Cook Rocks") -- *check.* Camera -- *not allowed.* Snacks -- *not permitted. People* magazine to read while waiting -- *check.* Sunscreen -- *check.* Water -- *check.* We were set.

A cab shuttled us from LAX to the Nokia Theater, where hundreds of fans eagerly lined up outside waiting for the big show while media interviewed the die-hard AI viewers who were lucky enough to have scored tickets. By 10 a.m., the line started to slowly creep along and we had become new best friends with the people in close proximity. We guarded each other's spot in line like a lion protecting its cubs from predators.

Noon -- *I'm starving*. Heather headed off to find a Starbucks for coffee and a muffin. Meanwhile, I unfolded my chair and flipped through *People* -- making sure not to miss the article about David Archuleta's alleged abusive father.

Over an hour later, Heather returns.

"I had to walk several blocks just to get a cup of coffee. There is nothing around here," she reports.

At 1 p.m., we were ushered into the Nokia Theater hallway like cattle being herded into a slaughterhouse. Mobs of ticket holders squished into the confined space and plopped down on the carpeted floor. It was hot and crowded. There was no food or water available. We had hours more to wait. I felt like a CAFO-raised cow, with barely enough room to turn around. If I were a meat-eater, after this experience, I would opt to eat only free-range meat.

By 3 p.m., we were seated inside the theater. The *American Idol* signs lit up the stage. I squeezed Heather's arm, "Can you believe we are here?" After watching it on TV for so many months, it felt surreal. I truly was the golden ticket holder.

Those of us who endured the daylong wait only filled a small portion of the 7,100-seat theater. We were treated to a full show rehearsal before the live event was to begin.

The stage manager gave us the run down. We could either watch the rehearsal or head to the snack bar for some (much-needed) food. Once Ryan Seacrest (adorable) took the stage, I wasn't about to leave my seat. *Damn, why hadn't I opted to upgrade to a camera phone?*

I'm sure Brian Dunkleman, who co-hosted with Ryan on season one, was kicking himself for quitting the multi-million dollar enterprise. Allegedly, he left the show because the contestants were treated poorly and he wanted to focus on his "acting" career. Since then his biggest credit has been his appearance on *Celebrity Fit Club*. Meanwhile, according to the Hollywood Reporter, Ryan just got a raise to $15 million a year.

On stage, David and David took turns wooing us with ballads. Much to my amazement, they sounded much better live. Their abilities blew me away. Goosebumps covered my skin.

Like a teenager at a rock concert, I waved my sign high in the air and screamed when David Cook came on stage. By the end of the rehearsal, with Ruben Studdard belting out a tune, Heather and I dashed to the snack bar in hopes of downing some morsel of a meal before the live show. We waited in line with the throngs of people who were also starving.

Just as I stepped up to the counter, the woman informed me that they were completely sold out of food -- yet apparently they still had plenty of beer available. A toxic combination. I was outraged, but it was of no use. The scene suddenly took on the air of the great depression, with long lines of people begging for food.

The theater filled up as the live show was about to begin. The noise was overwhelming. The judges took their seats to the applause of screaming fans. Ryan came out dressed to impress. The show began.

Yes, it might just be a TV show, but it was also an iconic piece of American history that I got to witness in person -- akin to *American Bandstand*. An experience I would not soon forget.

This year was a far cry from last year's AI finale. Sitting in our dingy hotel room, it was 9:45 p.m. and Katelyn nearly made it to the end. Her eyes were heavy and she struggled to stay awake as Kris Allen was announced as the victor.

With a TV at my disposal, I flipped channels and took in as much as I could. Making up for lost TV time, I tuned into *Make Me a Supermodel*, *Something About Mary*, *Medical Mysteries*, *I Shouldn't Be Alive*...you know, quality TV.

By 1 a.m., I was still glued to the screen. I flipped through the channels one last time and found *Oprah*. Oh, how I had missed her. Her topic of the day, ironically -- how to live without technology. For her "What Can You Live Without? Challenge," families had agreed to give up technology for one to two weeks. I had been living without it for a month. She explained that by giving up technology people could find meaning in their lives. "This week. It's all about disconnecting from all the stuff and all the technology and reconnecting as a family," Oprah said.

Finally, someone else gets it. Now maybe it wouldn't be so awkward to tell people about my journey. After all, *Oprah* has an estimated 42 million viewers. Surely, someone I know would see the show and appreciate what my family and I were putting ourselves through.

The biggest difference between me and the people on *Oprah*? The social ramifications. Her guests had it easy because (I assume) they told their friends, "I can't e-mail you this week because I'm going to be on *Oprah*, talking about not having e-mail." Whereas, my situation was much more difficult and showed the true test of where our social conscious lies and to what degree we are judged based on our connectedness through technology. It was inevitably awkward, embarrassing and inconvenient to admit that I was living a life unplugged.

Day 32: "Did you see *Oprah* yesterday? It's all about people giving up technology -- exactly what we're doing," I said to Marie. "But Oprah explains it much better than I do. You have to watch it, then maybe you'll really understand why we're doing this."

"I heard about it, but haven't had a chance to watch it. I Tivoed it. I'll watch when I can. I understand what you're doing. I get it...kind of."

Maybe I couldn't get through to my beloved cousin-in-law, but surely Oprah would. She was the heavy hitter I needed to get my point across.

In Paulo Coelho's book, *The Alchemist*, he talks about "omens" and how "when you want something, all the universe inspires in helping you achieve it." The fact that I was living a life unplugged -- without a TV in our home -- and just happened to book a hotel room at the last minute and just happened to be up at 1 a.m. watching said TV and just happened to come across *Oprah* who just happened to be talking about the exact thing I was experiencing...that's what I call an omen. That reassured me that the sacrifices I was making would be worth it.

***

As Jeff walked in the door, home from his trip to Japan, I grabbed his hand and walked him to the garage.

"Close your eyes."

I lead him down the step into the garage and announced, "OK, now open."

His eyes blinked open as he surveyed the well-organized, uncluttered garage.

“Wow! This looks great. It must have taken you forever. Thank you," he said.

As an anniversary gift, I spent the week he was away throwing out the old crap and bringing in the new storage boxes and shelves. Assembling, categorizing and tossing, to arrive at what now looks like a civilized garage, minus the Goodwill pile heaped in the middle. We would be selling what we could at our garage sale and donate the rest to those in need, including two of Jeff's shirts that I set aside to give to a homeless man I'd seen standing on the corner lately.

Day 33: With our map and a book about national parks, we headed off to meet our friends Don and Macy for a weekend of camping at Pinnacles State Park. We estimated the drive to take about an hour. About 40 minutes into the drive, we passed by a "Pinnacles" sign. We wavered, but pushed on -- knowing that there are two entrances. *Two* hours later, we arrived at the second entrance with a sign that informed us, "Day use only…no camping." Damn, wrong entrance. Jeff was steaming mad. Another two-hour drive back to reach the campground.

"We could've driven to Yosemite," I commented. My thoughts drifted back to my earlier comment, "I'm glad we're camping close by." Ha! How I longed for my iPhone with Yahoo Maps and GPS.

Knowing we'd be close to three hours late and it was nearly dinner, Jeff suggested we call Macy to update her on our arrival. Not an easy task without a cell phone. We drove on, keeping our eyes peeled for a payphone. We spotted one at a gas station. I picked up the receiver…no dial tone. I tried the second one. A dial tone. I put in my thirty-five cents and looked up Macy's cell phone number in my organizer. Damn, her cell phone number was a long distance number. Something she hadn't changed upon moving to the Bay Area two years ago. I plunked my money in and dialed anyway. The dial tone went dead. *Sigh. Do any pay phones work anymore?*

In 2007, AT&T announced its intention to withdrawal from the pay phone business by the end of 2008, due to a decline in demand as a result of wireless communication devices. The only place you might find pay phones anymore...prison, where inmates are allegedly being charged excessively high rates, which can make keeping in touch with their families difficult, leading to higher recidivism rates.

Business Insider listed pay phones as one the "21 Things That Became Obsolete This Decade," along with movie rental stores, getting film developed, CDs, maps, phone books, newspaper classifieds, landlines, and record stores. I hope this report in not accurate because without technology, my family has become dependent on these antiquated items.

Day 36: Jeff is treating the family to a night out at...Double D's. Lucky me. Not what I pictured. There were no large-breasted waitresses pushing Amstel Light. The menu did not consist of only Buffalo wings and fried mozzarella. As we slid into a booth, the fourteen TVs blaring overwhelmed me. An Amish person would be spun into an anxiety attack in a place like this.

The LA Lakers took up nearly every screen as they vied for a place as champions. Jeff was glued. Not exactly what I would call a quality family dinner. After wolfing down my spinach salad, Katelyn and I escaped the madness and headed home, thankful that we took two cars. Jeff stayed behind to keep an eye on the action (The Lakers, not the waitresses).

Although I'd rather have Jeff home with me, I was thankful that he wasn't a sports fanatic like some of my friend's husbands who spend the majority of their time with the game on (football, baseball, basketball), beer and remote in hand...for hours every weekend. The wives are what I call "sports faux-widows." Their husbands will never unplug. They NEED their sports, more than their wife sometimes, but what they don't realize, is that they are the couples that would benefit the most from taking a break from the TV.

Parenting Magazine's "Ask the Mom Squad" contributor Dr. Macy Berman recently gave advice to a reader on this particular topic. "While this can be healthy in small doses, he shouldn't be so fixated on the TV that he ignores you or the kids," she says. "Make your feelings clear by saying something like 'it hurts me when you're glued to the screen for hours on end, and it makes me feel as though you're choosing the game over me and the kids -- we want your attention, too'." She urges women, *not* to have this conversation *while* he's watching TV. Good luck, ladies!

Day 39: Katelyn and I spent the afternoon snuggled up in my patio lounge chair reading books. As I tilted back my zero gravity chair, I felt the warm summer breeze and heard birds chirping harmoniously in the tree above me. Katelyn hopped up on my lap and snuggled in for story time. After plowing through her books, I just laid there, arms wrapped comfortably around my darling daughter -- soaking in the moment. Katelyn broke the silence.

"Are we just going to sit here? This is boring," she said squirming on my lap. Startled back to reality, I replied, "I was just enjoying my time with you." I was not missing TV. Having a TV had given me an excuse to not spend quality time with Katelyn. Without a TV, I was forced to enjoy my daughter, uninterrupted. Getting rid of the TV gave me back valuable time needed to tackle my never-ending to-do list and still have time to just play with Katelyn. I also had time to read. It's shocking how quickly a book can be read when it is made a priority. Upon finishing a book in one day, Jeff asked how I was able to read it so quickly.

"It takes me forever to read a book," Jeff said.

Normally I would agree. Mostly because when I was reading, I was also simultaneously thinking about tomorrow's to-do list, planning our meals for the week, wondering when I'd have time to get to the pile of laundry that needed to be put away. With technology, we get used to our brains being bombarded with information...music, commercials, e-mails, weather, phone calls, news...it's a constant buzz of activity. Quieting the mind is not done easily. Giving up my technology has allowed me to stop and appreciate what I'm doing at the moment, which I have learned, from reading Eckhart Tolle's *The Power of Now,* is the key to "freedom."

He advises, to "Use your senses fully. Be where you are. Look around. Just look, don't interpret. See the light, shapes, colors, and textures. Be aware of the silent presence of each thing. Be aware of the space that allows everything to be. Listen to the sounds: don't judge them. Listen to the silence underneath the sounds. Touch something, anything (*including your husband?*). And feel and acknowledge its Being. Observe the rhythm of your breathing; feel the air flowing in and out, feel the life energy inside your body. Follow the 'isness' of all things. Move deeply into the Now."

I agree that to find peace and happiness, we have to unplug, disconnect, and turn it off. First I cleaned my garage, now I am cleaning my mind of clutter. Reading now allows me to enjoy some peace and quiet, both externally and internally. And unlike television where you can walk away with a feeling of time wasted, reading provides me with a much-needed sense of accomplishment. Even finishing one chapter makes me feel like I've achieved something. I'm that much smarter or better informed than I was before I read it. I would gladly brag that, "I read three books this week." But how often would we say, "I watched 15 hours of TV this week"? I can sit with my family and put my book down at anytime to interact. No waiting until a commercial or to finish e-mail.

Katelyn and I have been spending our TV-free afternoons outside at the picnic table. She paints, I read. I frequently look up from my book and comment on her "very creative use of shaving cream, glitter and glue."

Previous to unplugging, I would have my laptop open and my typical response to her request for me to gaze at her masterpiece would have been, "Just a second, let me just finish this one e-mail." Now I'm available, interested, and attentive. But also doing my own thing.

***

Chatting on my iPhone while composing e-mail and watching the *Today Show*, all at the same time, was normal for me before we unplugged. Ah, technology and the "joy" of multitasking. Technology allows us to get more done in a shorter amount of time, right? The interesting and ironic thing is that without technology, I'm finding that I feel less stressed and more productive. Unplugging has forced me to focus on doing one thing at a time. I am therefore more aware of what I'm doing -- giving it my full attention. My time is less frazzled and wasteful. Things have slowed down and I do things with more intent, rather than rushing. I'm more focused.

In his book, *The Guinea Pig Diaries: My Life as an Experiment*, author A.J. Jacobs says, “In one sense, task-juggling makes me feel energized, fulfilled, like I’m living three lives in the space of one. But I also know I’m overloading my circuits. I can’t think straight anymore.” He spends a month trying to “unitask,” which includes tying himself to his desk chair with an extension cord so he can stay put to focus on his work.

***

“I think we should buy the Ariel invitations since you’re birthday party is going to be at a pool,” I said to Katelyn as we perused the party favors at Party City.

“I don’t know, Mommy. I really like the ones with all the princesses,” she contemplated. “But Ariel is a mermaid and lives in the water, so I guess we should go with those. Plus, she is my favorite princess...this year.”

Buying party invitations for Katelyn’s birthday is a first. Every other year, I sent out Evites. This year they would be mailed. As I began filling them out, I figured I should include the address of the pool where we were going to have the party. No Internet to look it up. I had to dish out another dollar to call 411. The operator informed me that they didn't have the phone number listed (seems to be the norm with Comcast information). I opted for the number to parks and recreation. Voice-mail. Unsure if it was the right number, I left a message with little hopes of a return call. I dialed the number again the next day without luck. After a few more wrong numbers and voice-mails, I felt frustrated and longed for the ease of the Internet. The invitations had to go out. As I was contemplating my next move, I received a call back from the right person with the right information. I missed having instant access to information. We'll see how the RSVPs go without the Evites. People will have to actually pick up the phone and call to RSVP.

Day 40: Upon awakening, I rolled over and put my arm around Jeff and snuggled into his bare chest.

"What time did you get home last night," I asked sleepily.

"After eleven," he groaned.

"Did you go to Starbucks to do your work?"

“No. Double D's."

I lifted my head and looked at him, "Double D's? They have Internet access?"

"Well, sort of. If I sit against the wall near the manager's office I can get on his network -- a tip from the waitress."

***

I finally spotted him -- the homeless guy I've been looking for. I took a different route home and there he was sweating in the blazing sun on the meridian. The bag of used clothes I had been carrying around in my car for three weeks was at home. Quickly, I rushed home to get the goods. Along the way, I stopped quickly at Safeway to grab some water and sunscreen -- he looked like he needed it. I threw in a couple of Odwalla bars and whizzed through the self-checkout. As I sped down the street, a police car was nearby. *Oh, watch the speed.* I approached the corner and alas -- he was gone. *Damn it!* That's what I was afraid of. At a red light, I scanned the neighborhood. Maybe he was taking a leak at the gas station or enjoying some air conditioning at the donut shop. I sat there searching…and then I spotted him! He was crossing a parking lot across the street. I had to circle back. As I made my second U-turn, he took his position on his island, like a bird perched on a park bench waiting to be fed. I was stalking a homeless man. *Oh, dear!*

Once he was in place, holding his "Homeless -- Any help" sign, I positioned myself so that I could drive up and hand him my care package. The light turned red, there were two cars in front of me as I stopped near the meridian. I nervously waved him over to my car and handed him the bag. His face was sunburned, his long hair sweaty.

"What's in here?" he said rummaging through the bag.

"Some sunscreen, water, food, hat...," I was cut off mid-sentence.

"I have too much stuff to carry already," he grumbled, handing me back the loot and walking away.

"Don't you even want water?" I yelled after him, feeling rejected.

He walked back to his perch. As the light turned green, I slowed as I passed him and offered, "How about the hat and granola bars?"

"No, thanks," he waived.

What? Had my desperate attempt to help someone "in need" just been turned down? I guess Silicon Valley homeless are able to be choosey. I felt so dejected.

Upon arriving home, our gardeners were trimming the trees in the yard. Their English wasn't good and my Spanish needed some brushing up, but in my moment of wanting to help someone, I offered them an unopened bag of Trader Joe's peanut butter and chocolate chip cookies, which they gladly took. Good to know someone appreciated my efforts. After all, this was a big sacrifice; I don't willingly part with peanut butter.

***

It's been more than a month since we unplugged. I decided to get out of the house and grab a cup of tea at my usual hang out, the Los Gatos Coffee Roasting Company. Without my laptop slung over my shoulder, I felt out of place. My regular seat at the coffee bar, wedged between two other laptop users, was exchanged for a seat in the sun on a bench outside with the dog owners. My how technology affects every aspect -- even where we sit. I enjoyed frequenting this office away from home – buzzing with activity with small round coffee tables filled with casual businessmen and women frantically typing at their keyboards like birds pecking seeds.

There was always an interesting conversation to eavesdrop on...chatting about the latest technology gadgets, Web applications, Facebook follies, and many of the sixty-five thousand iPhone apps. Throw in the occasional screaming kids and sleep-deprived Silicon Valley moms commiserating over the lack of assistance from their overly worked husbands, and the environment always was an entertaining place to get work done. Somehow the vibe at the noisy coffee shop was more conducive to creativity than my peaceful, undistracted office at home. I dreaded being alone…in silence. As I settled on the bench, cup of iced tea in hand, I reached into my now large purse and grabbed the latest book I was reading, and enjoyed the fresh air and friendly dogs that surrounded me.

The *San Francisco Chronicle* recently reported that Sal Bednarz, owner of Actual Café in North Oakland, was urging customers to "leave their laptops at home and actually speak to each other" -- at least for one weekend. Customers will be asked to unplug and converse at communal tables instead. *Way to go Sal!*

Day 44: As I embarked on a new hiking trail, I had some trepidation since it was not a populated path. I'm not comfortable hiking alone in the wilderness. Fear of weirdoes, fear of mountain lions. *Is it media hype or a real threat?* Watching a TV special about the woman that got mauled by a mountain lion in Whiting Ranch, Orange County didn't help.

I begin my journey away from civilization. The rocky uphill terrain of the Limekiln Trail proved challenging. Maybe I should've worn my hiking boots rather than running shoes. At the moment, I was thankful for not having my iPod. I wanted to be aware of my (possibly threatening) surroundings. I felt some sense of comfort as I tightly gripped my pepper spray -- my finger poised and ready to fire. As I walked up the wooded trail, the swish of my Camel back occasionally sounded like footsteps following me and I stopped dead in my tracks to look behind me to make sure I was indeed alone.

Listening to birds singing their songs, watching squirrels busily scamper across the path, smelling the earthy scent of nature -- I was thankful to be unplugged. Without TV, I have been forced from the gym with its rows of treadmills, stationery bikes, ellipticals -- each with it's own TV screen. Like hamsters in spinning wheels, gym-goers run in place while being rewarded with their favorite TV shows. After twenty minutes of trudging up a steep part of the wooded trail, I felt my body start to relax.

My trigger finger wasn't quite as tense. I began to enjoy my surroundings and embrace my solitude. I'd never been on a hike that far out by myself. Then suddenly the roar of an airplane sounded frighteningly similar to that of a mountain lion and I stood sill and quickly surveyed my surroundings for eminent danger as my heart raced. *No, just a plane.* I breathed a sigh of relief. After an hour round-trip trek, I returned to my car feeling as if I'd conquered a monumental undertaking. For now, the fear of the forest had shrunk just a bit.

Day 45: My mom friends and I gathered around a picnic table shaded by some Redwood trees as the little ones frolicked on the play structure nearby.

Lola says to me, "We're updating the Mom's Club roster. What's your cell phone number?"

"Uh, I'm not really using my cell phone anymore," I admit.

"Oh, that too. Next thing you know, I'll have to be calling you on a tin can with some string," she jokes.

With the electromagnetic fields from cells phone being accused of causing potential health risks such as brain cancer, low sperm count, Alzheimer's, and salivary gland tumors, a tin can could be safer.

Day 48: With free tickets to a minor league baseball game, the San Jose Giants, my friend Macy called and said, "Why don't we just meet there? Call me on my cell phone when you arrive and I'll meet you with the tickets."

I paused awkwardly -- too embarrassed to tell her I don't *use* my cell phone anymore.

"Or, why don't we drop the tickets off at your house on our way," she finally broke the silence. *Whew.*

"That would be great, if you don't mind. Then we can just follow you guys to the game (since we don't have Yahoo Maps!)," I said.

On the way to the game, Macy's husband Don punches it through a yellow light.

"Don't lose them," I yell to Jeff. "We don't know how to get there and can't call them on our phone if we get lost."

Upon successfully arriving at the stadium, we sat in the crowded bleachers rooting for our local team. We could not distinguish what exactly the mascot was. Macy turns to the guy behind us to inquire if he knows.

"No, but I can look it up," he said, tapping "San Jose Giants mascot" in Google on his iPhone. He holds up his iPhone for us to see, "It's a Giant." Suddenly his phone rings. It's his wife...calling from the snack bar line asking if he wants mustard on his corndog.

Day 49: Jeff and I did not spend our ninth wedding anniversary at a romantic Napa bed-and-breakfast. Nor did we spend it cozying up to a fire. We celebrated it at Double D's, of course. Being the mom of an unplugged child, I had to be creative to make sure Katelyn wasn't feeling utterly deprived, so she was off to a movie night with a friend that was also a huge *Camp Rock* fan.

The night before our date, Jeff casually mentions, "The Lakers made the playoffs. They are playing tomorrow night. You, by chance, wouldn't want to go watch it would you?"

Of course, I lovingly reply, "Nothing would make me happier (than spending our anniversary at a loud bar crowded with drunk, sweaty men chugging bear and yelling, 'Make the shot, Kobe!')."

He wasn't off the hook quite that easily. Jeff does have a way of getting me to approve his "guy time." It's slyly manipulative. I'm a sucker for the "You're such a cool wife" quote when I let him do something that I really don't approve of (like spending half of Thanksgiving day doing a mountain bike race). Although, I'm pretty sneaky about getting my way too. If I don't want him to do something, I usually pull the old, "I'd rather have you spend time with me, but if you'd rather not, I understand."

For our anniversary, we came to a compromise -- an hour of wine tasting at swanky new hot spot in Los Gatos and an hour to catch the end of the Lakers game at the sports bar. I had reluctantly agreed, but the ironic thing is that after getting my fill of romantic time, I was more than open to accompanying Jeff to his home-away-from-home. And, I decided to make the most of it. Watching the game, I yelled, I clapped, I drank...and snuck in a few kisses.

He was glowing about being able to share his interest in the NBA with his adored wifey. This really was quality time. And I had a blast. Life without a TV was more fun. It forced us to get out of the house and try new things -- to spend more time with each other, and friends.

While at the bar, I discovered that Double D's is not named after huge cleavage. The owner happens to have a first and last name that starts with D. Convenient.

Day 50: "Ya know, Bailey is a high School student and I did meet her at the *Apple* Store," I said to Jeff while folding the laundry. "I notice she has her iPhone with her at all times while babysitting Katelyn. Three things concern me about that. One, that she'll neglect Katelyn by texting her friends. Two, that she'll chat on the phone with her friends and talk about inappropriate subjects. Three, she'll let Katelyn (who is supposed to be unplugged) play games or watch videos on her iPhone."

Jeff replied, "She seems responsible. Let's just cross that bridge when we need to."

***

Parked in the driveway, Jeff is squished into the front seat of the car, indulging in his cup of mint chocolate chip ice cream while listening to the Lakers game on the radio -- since the radio in the house doesn't get that station.

In my fuzzy slippers and robe, I go out into the dark of the night to join Jeff engrossed in the game. Kind of romantic -- naked under my robe -- *maybe we could get busy in the back seat since Katelyn's asleep in bed?* We sit side-by-side, listening attentively to the announcer's play-by-play interpretation of the action. I envisioned Kobe dripping with sweat as he dunks yet another one in the basket and the fans wildly cheer him on. This must have been what it was like for our grandparents, gathered around the living room radio listening to a nightly program in the 1930s.

***

Oh shit (and I don't use that word lightly), I thought I was swimming along just fine without online banking or Quicken, using my checkbook register to track expenses instead. When I was at Safeway purchasing my quart of chocolate peanut butter ice cream (something no freezer should be without), my heart sank as I noticed my bankcard was missing. My mind raced to pinpoint the last transaction. The ATM!

Since I can't check expenses online, I've been printing out the "last 10 transactions" at the ATM to try to keep some sense of how much is in my account (and racking up a fee every time).

The ATM ate my card. This is not the first time. Why is it that my ears just don't hear the *beep, bee*p warning to take your card before it gets swallowed by the ATM. Can't they have a siren with red flashing lights and a voice that says, "Take your card now or you'll be living in purchasing hell for the next five to seven business days while waiting for your replacement card...since writing checks is a pain in the ass."

***

Day 51: Time to cross the bridge. As I sat in my Office, I heard, "Nobody's Perfect" playing. I peered out the office door to see Katelyn watching the *Hannah Montana* video on Bailey's iPhone. Not wanting to reprimand Bailey in front of Katelyn, I let it go. Ten minutes later, they were still on the porch engrossed in a YouTube video, so I politely suggested they ride bikes. Bailey put her phone away.

***

Jeff is in Japan on business. Since I can't call him on his cell phone, I had to call his hotel. I put it on speaker so Katelyn could hear. The receptionist answered in Japanese. I then asked for Jeff's room number. Katelyn said, "How did you know what he said?" Once again mommy is impressive.

Jeff answered and the first thing he said was, "I was on Google and it's in Japanese."

I miss my iPhone, especially with Jeff out of town. Normally, we'd use the “flight tracker” app to follow Daddy as he jets across the globe, and the time converter so Katelyn can know what time of day it is for Daddy.

As Jeff was leaving for the trip, I tossed him one of the disposable cameras and asked him to take photos.

Upon returning from his trip, I asked, "Did you take picture to show Katelyn what Japan looks like?"

"Some, but it wasn't easy." Jeff said, unpacking his suitcase. "I was so embarrassed that my colleagues would see my disposable camera. The Japanese are very into their high-technology digital cameras. I had to hide my crappy disposable."

Day 59: It’s time to undo the last two months of too much wine and chocolate (hey, something had to replace my TV habit). My new personal trainer Brian promised to have me "sweating buckets." After doing my assessment, he prepared a workout routine that guarantees a six-pack in about three months.

"You’ll get results if you eat what I tell you to and stick to the workout plan," he threatens.

"I prefer running outside for my cardio since the weather is nice," I say politely.

Like a drill sergeant, he replies, "While you're on my program, I want you to workout on the cardio machines at the gym where I can control the variables, but feel free to run outside on your 'rest day'."

He has no idea that he might as well have just sentenced me to twenty lashes because for me the worst torture is being stuck indoors on a treadmill with no iPod or TV to keep me entertained. Four days of week...I began to reconsider. Then I decided to take on the challenge just to see if I could do it.

My thoughts drift back to my marathon training days when I made the dreadful mistake of forgetting my iPod on a 16-mile run day. A few miles in, I turned to Jeff and whined and complained that I just couldn't do it without my music. "This fucking sucks," I finally spewed.

The thought of nearly three hours of running with only the sounds of nature to occupy me nearly brought me to tears. Somehow, I survived, but the memory still haunts me.

***

Katelyn's school was holding a fundraiser at McDonald's. Kids were running around screaming, laughing, balloons tied to wrists, ice cream cones dripping. I took a seat at a booth next to my friend Diedre.

She asks, "Did you hear we got robbed? They took all of our computers AND back-up drives, and our iPods."

"Oh, no!" I said. "That's my worst nightmare. They took your back up drives too?"

"Yeah, you think to prepare for a computer crash, but not robbery," she said. "The worst part is that most of our pictures of the kids were on there, from when they were babies. We lost all of those. So make sure you keep copies of your photos someplace safe."

"I burned all of our photos to CDs and am going to put them in a fireproof safe," I told her.

She replies with grace, "We're just happy that no one was home when it happened and that no one got hurt."

After hearing Diedre's story, I decided to put all of our technology in a plastic bin in the garage marked "tampons." Even when we do plug back in, I'll definitely keep my back up hard drives someone safe -- rather than keeping it conveniently stored in my office near my computer.

Month Three: iTunes Makes the World Go Round

"Technology…is a queer thing. It brings you great gifts with one hand, and it stabs you in the back with the other."
--Carrie Snow, Comedian

Day 60: Stopped at a red light, windows down, warm sun streaming into the car, Jeff says, "Is that your phone?"

"No," I answer. "I don't hear it ringing."

"Doesn't it have a bird chirp for a ring?"

I start cracking up, "It's not my phone. That's a *real* bird you hear!"

***

Our morning ritual now consists of me reading the *San Francisco Chronicle* at breakfast -- catching the headlines, checking out the daily weather forecast, reading my horoscope. Katelyn begs me to read the comics. Garfield is her favorite because she's seen the movie, of course. She likes Peanuts too, even though she has never quite taken to the *Charlie Brown* TV specials. They are so nostalgic for me. On Sundays, when the funnies are in color, we lie in bed and snuggle up and read them together. Not having a TV hasn't totally cut off our exposure to advertising. Katelyn memorizes the Toys "R" Us ads in the newspaper and wants all the girly stuff they are selling.

***

After spending yesterday afternoon crisscrossing the neighborhood hanging up our "Garage Sale" signs, I was disappointed to see that most of them were down today. I'm an obvious novice at this garage sale thing. *Were they knocked down by the wind? Did homeowners take them down? Did competitors steal them? Maybe I should have hung them on the morning of the sale. Do I need to hang more signs? How am I going to get anyone to come? I miss Craigslist.*

Katelyn has a plan. "We get up at six in the morning, you know, at sunset…"

"You mean sunrise," I gently correct her.

"Yeah, yeah. That's right, sunrise," she shrugs it off. "So we'll get up at six and hang the signs."

Day 63: Jeff and I were leaving Saratoga after a lunch date while Katelyn was in school. We were enjoying a ride in our new Prius, when I heard a loud thump.

"What was that?" I turned around to see some young girls looking at us and laughing.

Jeff said, "I think they just threw a rock at our car!"

"What? Slow down," I shouted.

As the car rolled to a stop, I jumped out and chased after the girls. How dare they scratch my new car that took me three hours to purchase while Katelyn whined "I wanna go hooooome!" every time the salesman left the room.

I pulled a Bruce Jenner and sprinted after them, yelling, "Stop!"

Two of them surprisingly did. One kept running. My adrenaline was pumping.

"Did you just throw a rock at my car?" I panted.

"No," they denied.

"I heard it," I insisted. "That's a *new* car. You don't go throwing rocks at people or their property," I yelled. "How old are you anyway?"

"Thirteen," one girl mumbled, looking at the ground.

"You should know better," I scolded. "I want your names. This is totally unacceptable behavior."

"Well, it wasn't a rock," one girl piped up. "It was a pod from the tree."

"I don't care what it was. You do not throw things at people's cars."

Then I heard myself saying, "I'm going to call the police."

"No! Please don't," they both begged as tears welled up in their eyes. "We'll make it up to you. What can we do?"

Jeff walked up after parking the car and assessing the damage (none).

"Do you have your cell phone?" I said to Jeff. "I think we should call the police." I knew he didn't have his phone and I wouldn't really call the police, but I thought the girls needed to learn a lesson.

Jeff replied, "There was no damage, but what were you thinking?"

I added, "You made a bad decision that can lead to taking the wrong path in life and, believe me, you don't want to go that route. I know kids who have been there, and it's not pretty. You need to think about what you are doing and the consequences. It's not worth it! Don't ruin your lives. Make smart choices. We won't call the police, but I hope you make better decisions next time and learn from this."

As we walked away, Jeff said, "They're just bored."

"That's what will get them into trouble," I worried.

Walking back to the car, I realized just how important a cell phone would be in case of a real emergency.

Day 64: We got our phone bill today. We racked up thirty-five dollars in 411 calls. And I only use it when I have to -- if I can't find the number in the phonebook. Looks like, in this case, the Internet would save us money by allowing us to look up phone numbers online.

I decided to investigate why the bill was so high. I called the Comcast Billing department and spoke to four different people. Each of them put me on hold for a lengthy amount of time until one guy told me that 411 calls cost $1.50 a minute.

"Per minute?" I asked. "So, if they take more than a minute to give me the phone number then they can charge me three dollars?"

"Guess so," he said. "But let me just verify that information."

The agent dialed another agent, unaware that I was still on the line. He politely inquired about the cost of 411 calls. They both addressed each other as "sir" and were extremely polite to one another. The new agent says that it's ninety-nine cents per call. The first agent gets on the phone with me to tell me the correct price.

"It's ninety-nine cents per *minute*," he says. The second agent is still on the line and speaks up, "It's per call, not per minute."

"How many 411 calls did I make during my last billing cycle? If I got charged, that would mean that I called 411 nearly 35 times in one month. I am sure that I did not call that many times," I tell the agent.

"I'm sorry, I don't have that information," he informs me.

"Seems like it should be listed on the bill," I say.

"You'll have to call the phone department," he replies.

I call the other department and am put on hold for nearly 10 minutes. When someone finally answers the phone, she tells me, "You can look up your specific account information online at comast.com."

"I don't have Internet access," I tell her.

She puts me on hold and returns. "It's rare that we have a customer without Internet access. You'll have to go to your local Comcast office to get that information," she says.

Shocking! So I have to get in my car and drive down to the office to find out how much I'm paying for 411 calls all because I don't have Internet access.

Day 65: Bailey is babysitting. I was supposed to go to my sewing class, but it got cancelled, so I decided to get my fix for entertainment and go to a movie…by myself. I know, pathetic. But I'm feeling a need since I don't get my TV fix anymore.

As I leave, I tell Bailey that Jeff should be home by 6:30 p.m. Since Bailey doesn't have a driver's license, her mom picks her up at my place when she's done. If I'm late getting home, Bailey's mom has to wait. I tried calling Jeff a couple of times at work, but got his voice-mail.

At the theater, I bought my ticket and found out that it's not over until 6:45 p.m. I need to make sure Jeff's home by 6:30. I hand my ticket to the ticket-taker, an elderly lady with gray hair and a wrinkled smile, and ask, "Is there a payphone around here?"

"No, it's a shame. They took it out," she replies, shaking her head.

I walk to the information counter and ask to use their phone for a local call. The teenage employees are more than happy to oblige, but look at me as though I'm a senior citizen. I feel so not hip. One phone doesn't work, the other comes unplugged and has to be reset. Meanwhile, I'm missing the start of the movie. Finally, a dial tone. I call Jeff. He answers.

"Can you be home by 6:30?" I ask.

"No," I have a meeting.

"How about seven?"

"If my meeting is done by then," he says.

"OK. Bailey is babysitting. I'll tell her you'll be home by seven. Don't be late!"

I call home and get the answering machine, "Bailey, it's Sharael. Hello? Pick up. I'll call your cell."

The movie theater employees are standing behind the counter chatting about their boyfriends and paying no attention to the several calls I'm making on their business phone.

I call Bailey's cell phone…voicemail. "Bailey, it's Sharael. Jeff won't be home until seven. If that's not OK for your mom to pick you up later, call me on my cell phone," I say desperately. After all, I consider it an emergency.

I couldn't totally relax during the movie because I was worried my cell phone would ring and annoy those sitting around me. *Did Bailey get the message? Would her mom have to wait around for a half-hour longer than she thought?*

I rush out of the theater as the final credits roll. Jeff and I pull up to the house at the same time. Bailey got the message. Things were fine. Man, I miss the convenience of a cell phone, but am also thankful not to be at risk for "text thumb," "cell phone elbow," "Blackberry neck" or "computer vision syndrome." One young woman in New York even recently fell into an open manhole -- she didn't see it because she was texting!

Day 66: Without a TV, I have become more aware of how we spend our money. Advertising is a multi-billion dollar industry because it works. I have unknowingly fallen victim to purchasing things I don't need because of a convincing commercial.

Time to cut back on *Hannah Montana*, *Cinderella*, and *Camp Rock*. See a theme here? Disney. They really know how to get parents to over spend…constantly buying things their kids just don't need.

Believe it or not, there are several anti-Disney groups with members who accuse Disney of everything from the obvious (violent movies, degrading women) to the unthinkable (sweatshops, pedophiles). Yet millions of people still flock to Disneyland and Disney World every year, and leave with an armful of trinkets to take home with them.

It used to be that I would buy Katelyn brand items because she liked them. It made her happy. Why would I deny her yet another *Hannah Montana* doll when she only has twelve of them? I adore seeing her smiling face. But now I'm going to lean towards more generic items. They are less expensive and don't promote commercialism quite as much. If I buy into the brands, am I telling Katelyn that she needs those things to feel good, be cool, or fit in? As she gets older, will she be learning that she needs brands to be popular or to like herself and her life? I want her to feel good about herself whether she is wearing Juicy Couture jeans or Merinos jeans from Target. I want her to grow up with a sense of pride in who she is and how she treats others, not what she owns. Material items are just icing on the cake.

Day 67: Today was my first trainer-sanctioned workout. An hour on the treadmill. I spent the first 20 minutes thumbing through *People* magazine and 15 minutes flipping the pages of *Redbook*, leftovers on the gym's magazine rack. After that, I just stared at the blank TV screen attached to my treadmill. Grueling. It didn't help that the women next to me was chuckling at her Bravo show -- some makeover program. I sneak a peek. Not the same without the audio. Only 20 minutes left. The blinds are drawn over the window in front of me. I can't reach them for fear of rolling off the treadmill and onto my face. My treadmill neighbor lets out a slight laugh again. I glance over. Two guys are yelling at each other on the TV. I don't remember laughing out loud at many shows. Ugh, 15 more minutes. Maybe a 45-minute workout would be good enough. *Do I really need to do an entire hour? How the hell am I going to do this five days a week? Torcher.*

Day 68: Today was the big day...the garage sale. We hocked our "trash" to takers that saw it as treasurer. Cars crept by, trying to sneak a peek at what we had sprawled across our driveway...stuffed animals, clothes that Katelyn had outgrown, books that I was done reading, food storage containers that had been cluttering the cupboards, artwork that we were board of looking at, a rickety desk chair...the usual. Katelyn loved playing cashier. She had taken the time to put little blue dot stickers with prices on each object that she was willing to part with and gladly took people's money in exchange for her junk.

I was amazed at how much the garage sales "professionals" haggled down the prices from a few dollars to a few cents. In the end, they were saving me a trip to the Goodwill, so I took what I could get. A leather jacket, we had picked up in Italy for two hunderd dollars, sold for twenty. A bag of Katelyn's "designer" duds, went home with a new family for a mere ten bucks, about a dollar per item. A bin full of Seasame Street stuffed animals seemed to wave goodbye as they got stuffed into the trunk of a new car for only five dollars.

At the end of the day, we were left wtih nearly two hundred dollars in cash and a driveway cleared of clutter. Rather than pocket the money for ourselves, we decided, as a family to donate the money to a children's shelter down the street. The money from our garbage was enough to purchase more than a dozen backpacks full of school supplies for children in need. Everybody wins.

Day 69: Denise and I decided to take the kids for a picnic at Emma Prusch Farm in San Jose.

"So, where do you want to meet?" I ask.

"Oh, I guess just in the parking lot," Denise replied.

We arrive a few minutes late and I don't see her car, so we get out and take a seat at a picnic table. My cell phone rings -- it's Denise. I feel so utterly rude not answering it. I keep a close eye on the parking lot. *What if she was calling to say she couldn't make it?* A few moments later, they walk up.

"So, you're not using your cell phone either?" she inquires, responding to the voicemail she heard.

"No, just for emergencies," I admit. "Did you just call?" I lie. "I usually just leave my cell phone in my glove box."

"Yeah, I heard your message, but left you a message anyway, just in case you were checking," she said.

Over lunch of PB and J sandwiches and carrots sticks for her kids, and pink marshmallows and pretzels for Katelyn, I fill Denise in on our reason for unplugging.

"Do you find people are not contacting you as much because it takes extra effort?" she asks.

"Well, I don't know. If someone sent me an e-mail and then decided not to call to follow up, I'd never know about it since I'm not checking e-mail," I said.

The funny thing is that at the end of the day she asked what Katelyn would like for her birthday and I said I'd have to think about it and she said, "Oh, you can just e-mail me later."

"But I'm not on e-mail," I joked with her.

It's ingrained in our society to be always accessible and plugged in.

***

Before walking out the door to for Katelyn's kindergarten graduation -- complete with caps and gowns, Jeff pleads with me, "We just have to video tape her graduation."

"Yeah, OK," I agreed to cheat on our no technology rule.

I rushed to the garage and dug through our box of technology…no camera.

"It's not there," I say to Jeff, bummed but not completely freaked out.

Scrambling not to be late, I check a shelf with the camera box…it's inside! The battery is dead and the disk almost full. I plug it in and say to Jeff, "Technology is such a hassle."

It added a sense of stress to our morning. I semi-charged the battery and guiltily stuck it in my bag.

After dropping off Katelyn and Jeff at school, I drove to Safeway to buy some flowers for Katelyn. I asked if they have DVDs for my video camera…no luck. *Oh well.*

At school, I ask her teacher if they will be recording the graduation…surely they would.

"No, we don't have a school video camera," she said.

My heart sank. I was desperately looking for a way out…not wanting to cheat on my no-technology lifestyle. I was trying to stick by the rules and see if there was another way to get the video other than using my own camera. Also, I was unsure if my battery or DVD would run out halfway through.

My friend Denise sits down next to me in the folding plastic chair.

"Are you video taping this?" I ask desperately.

"Well, our camera got stolen, so my husband is going to use the video camera on his iPhone," she says.

Oh! I'm so jealous. I want to be one of the 40 million Apple iPhone owners.

"Do you think you could burn us a DVD? My video camera is not working well," I say.

"Sure!" she says.

Once the performance starts, I can't resist pulling out my flailing video camera -- I just couldn't miss it! About halfway through the first song, the DVD is full. I'm grateful that Denise is recording it (upon returning home -- the camera goes back in the garage). The kids sing two songs, do a dance routine to a *Hannah Montana* song (of course), and then walk across the stage to receive their diploma and say, "I love you Mommy and Daddy," into the microphone. A touching day that I will hold fondly in my memory.

***

"When and where are you meeting your friends for the bachelor party tomorrow night?" I ask Jeff suspiciously.

"I don't know. They haven't called me yet with the details and I tried calling, but don't have the right cell phone number," Jeff said. "I'm going to have to send an e-mail to find out what the plan in."

"No way! You can't. That's the whole point of unplugging," I said. "It's not supposed to be easy."

Jeff, for the first time since unplugging, has to face what life is really like without technology.

"But I bet Glen sent out an e-mail with the details and I didn't get it since I'm not on e-mail," he said, getting defensive.

"Well, your away message has our home number and says that you're not checking your e-mail," I say to him.

"No one reads the away message," Jeff huffs. "Everyone assumes it says you're on vacation."

"Does he have your cell number? Well, you're not supposed to use that either though. If he calls it and gets your voice-mail, he won't know to call the house. I think you need to change your voice-mail to say that if it is a personal call to try you at home," I lecture Jeff.

"No, most of my friends and family know to call me at home now," he said. "If I miss this bachelor party, it could really impact our friendship."

"I know how important it is to you," I said. "I deal with these types of dilemmas nearly daily. It's awkward and uncomfortable."

"I'm sure if I'm not there for dinner, he'll call me afterwards," Jeff says.

At 10 p.m., Jeff is fast asleep. The phone rings. It's Glen.

"Hey, Sharael. Is Jeff there?"

"He's asleep. He's been trying to get in touch with you because he didn't know where to go for the bachelor party," I tell him.

I can barely hear him because of the noise in the background.

Someone yells, "Get off the phone!"

"We're meeting at 6 p.m. for dinner tomorrow," he tells me. "I sent him an e-mail with the address."

"Oh, yeah. Well, he doesn't have e-mail access. Can I get the address? And what's your cell phone number -- the one he has doesn't work."

Jeff is in! *Did I just help facilitate my husband attending a bachelor party? What was I thinking?*

***

While shopping for Katelyn's birthday at Toys "R" Us, I saw that Leap Frog has come out with a kid's pretend PDA. The package states, "I just sent you a text."

Now that's something that a couple of months ago I would've bought and thought it was so cool, but now I just don't feel right promoting such a toy. *What kind of message is it sending to our kids -- promoting a plugged in, multi-tasking, distracted lifestyle at age six?*

***

I feel like Katelyn needs something meaningful to replace TV…a pet. She's always wanted a guinea pig. With her birthday coming up, it would be the perfect gift -- something to come home to after school that is entertaining yet soothing. A constant companion.

The only problem is Jeff. He has an overly sensitive sense of smell and is not keen on the idea of adding a guinea pig to our home.

"I don't want to come home to a house that smells like piss," Jeff said defensively.

"But she would love a pet," I pleaded.

"It's just going to stink up our whole place."

"I'll clean the cage every day," I laid the guilt trip on him. "I'm willing to do that to make Katelyn happy."

He still wasn't convinced. I decided to take Katelyn to the animal shelter just to look at the pet options. The Humane Society of Silicon Valley is a newly built twenty-five million dollar facility. It's like a pet hotel. No cages. Each dog has it's own room, own bed with a blanket and toy, and a nice patch of fake grass. Katelyn and I peer through the windows -- a pit bull, a black lab, a terrier, and the cutest, most pathetic, yet adorable, tiny Chihuahua named, of course, Hope. Good advertising. "I *hope* you'll adopt me," she seems to be saying as she looks up at us, trembling with excitement. She stares at us with her forlorn eyes. I'm not a fan of small dogs, but I could've stuffed her in my purse and taken her home with me right then and there.

We look at the cats. Stepping into the Cat House, the smell of cat food and litter boxes is overwhelming. Plus, Jeff is supposedly allergic to cats. I grew up with cats and love them. So soft and cuddly, but I think I'm now swaying towards Jeff's side…too smelly and I couldn't stand the cat hair getting all over everything.

Next stop, bunnies. So soft, but not really playful. There are two grey ones, a brown one and a white one. They would have to live outside, which isn't much fun. And they tend to scratch.

I grew up with lots of animals…dogs, cats, bunnies, a goat. I guess, as a kid, I didn't realize how much work it was to take care of so many pets.

"What about a cute parakeet?" I asked Katelyn as we stroked the velvet fur of the rabbits.

"That's no fun. You can't play with them," she frowned.

As we are leaving, I notice a brochure on the counter about foster homes for pets. I inquire about the program. A Humane Society volunteer informed me, "We offer dogs and kittens to a temporary home. We don't recommend dogs for families. They might not be safe for children, but the kittens are great. You keep them for two to six weeks."

"How do you sign up to be a foster home?" I asked.

"You have to do it online."

"Do you have a computer here that I can use to do that?"

The volunteered said I was welcome to use theirs.

That night I told Jeff about fostering some kittens. His response, "I just think Katelyn would get too attached to the kittens and would be broken hearted to give them away."

For now, pets are on hold.

Day 71: Brenda and I were going to take the kids to the Oakland Zoo, but it's supposed to be 95 degrees, so we opted to meet at the club to go for a swim -- a first for them. We were supposed to meet at 11 a.m., but they didn't show up. *I bet she tried calling me on my cell phone*, I think to myself. By noon -- still no show. One o'clock -- no sign of them. I used the phone at the front desk to call her at home. No answer. I leave a message.

"Hey, I just wanted to make sure you didn't go to the wrong club or get sick. We're at the club."

Katelyn and I stayed and enjoyed the sunny summer weather until 3 p.m. Katelyn was very disappointed that her friends didn't make it. She was looking forward to playing with them on her birthday. I called Brenda when I got home and her daughter answered. She informed me that her mom had been rushed to the hospital for kidney stones.

Day 73: Jeff isn't feeling well. Katelyn said, "Daddy has a sore throat like Wanda on the *Magic School Bus*."

"Where did you see that show?" I asked curiously.

"At kindergarten. We watch a lot of *Magic School Bus* at school and we learn a lot. That way the teacher doesn't have to teach us so much."

So glad we're paying an arm and a leg for private kindergarten -- wouldn't want the teacher to actually have to work for her salary or anything.

Day 74: Katelyn and I headed to the Oakland Zoo for a day of animal watching. On the way there, I noticed a lot of road construction and had a feeling we would probably be more than two hours late getting home that evening. I didn't want Jeff to worry, so I attempted to find a payphone to let him know the situation. I looked on the zoo map and saw a payphone icon, so we went in search to find it. No luck. I asked a janitor. After looking at the map, he said it must be near the entrance gate. No luck. I asked the cashier. He said it must be just a few yards away. He motioned to a security guard. "Can you show her to the payphone?"

"Oh, they took that out years ago. I guess the phone company didn't want to pay for it anymore," he replied, shrugging his shoulders.

"But it's on the map," I told him.

"I guess they haven't bothered to update the map," he said.

Poor Jeff would have to trust that we didn't intend to be late.

***

We have a house full of sickies. The phone rings…949 area code…it's for Jeff. His old lifeguard buddy is calling to let him know about a reunion in Laguna Beach. He said he got our away message on e-mail and decided to call instead. Nice of him to make the extra effort.

"If I'm telling you that I'll be sending you e-mails, will you check that account?" he asks Jeff.

"I can be reached by phone or regular mail," Jeff replies bravely.

Day 76: Katelyn's music teacher gave me the names of the two songs they will be learning in the class. She said, "You can just download them on iTunes if she wants to practice."

Well, I can't. So I went to the library and found a CD with one of the songs and had to drive halfway across town (and San Jose is a big town) to another library to get the other song. Mission accomplished without spending two dollar on iTunes.

While browsing the CD section, I decided it might be fun to expose Katelyn to various types of music (rather than the same Top 40 songs on the radio). I chose a selection of country, classic rock, cultural, and

French tunes. If I would not have unplugged, I don't think Katelyn would have been exposed to these new types of music. I would not have bought these CDs and would have been too lazy to check them out at the library. This is not only for Katelyn…I'm dying to listen to something other than Lady Gaga and Jason Mraz.

Day 80: When I started my journey of a TV-less life two months ago, I decided this would be the time to catch up on some much-needed reading. I bought 27 books. To date, I've already read six of them -- trying to average one book a week. It hasn't been very difficult to fit the time in -- while Katelyn's doing art projects or homework, while waiting for her at voice/swim/dance classes, while at the gym, while waiting at the doctor's office, after she's in bed.

Since I've been reading so much I thought it would be nice to have some people to discuss the books with, so I've decided to start my own book club. Of course, *I* will get to pick the books. Only problem, without e-mail or the Internet, how am I going to get members?

To solve this problem, I decided to make it a neighborhood book club for women. I *typed* up some flyers and dropped them in the mailboxes of my immediate neighbors. Katelyn thought it was a grand idea. She stood on her tippy-toes and slipped the notes into the mailboxes.

"Great, I'll kill two birds with one stone…get to know the neighbors, and discuss books I've read," I said holding Katelyn's hand as we strolled back to our house.

Later that evening, my neighbor Deidre called and left a message, "I'd love to join your club. How often are you meeting? How long do we have to read the books? I like your selection. Give me a call when you have a chance. Oh, this is Diedre, from across the street."

***

At the evening Famer's Market, Jeff had roasted chicken and rosemary potatoes for six dollars, Katelyn had a strawberry crepe for seven dollars, and I had a corn on the cob for two dollars. The Farmer's Market closed at 8 p.m. and I wanted to buy some fresh fruit. I ran home to get some more cash and got back just as they were putting things away. Just before closing, the prices dropped from two dollars to one dollar a pound for fresh organic fruit. They were practically giving the food away. One vendor offered four ears of corn for one dollar; another gave me a loaf of bread for two dollars off, after I bargained him down. What a find! I'll shop the Farmer's Market every week from now on.

This is just the sort of thing I would love to e-mail my friends about. Although, maybe I'm the only one who didn't know about it. I was always e-mailing friends about bargains (sale at Banana Republic), events (puppet show at the library) or get-togethers (Mom's Night Out at Santana Row).

When I was just out of college and living in San Francisco, I would regularly send a group e-mail called the "Partyline" to inform my friends (and anyone who wanted to join) of goings-on in the city…there was always something interesting to do in San Francisco.

Day 85: The new generation of iPhones came out recently -- now they include a video camera. I'm so jealous. I long to trade in my big, bulky purse and Day Runner for a sleek and slim, multi-functioning iPhone.

I remember being at the MacWorld trade show in 2007 when the first iPhone was announced. They only had two on display -- encased in what looked like bulletproof Plexiglas with staunch security guards protecting the priceless treasures. I sat enthralled watching the presentation of all of its capabilities -- couldn't wait to get my hands on one, but wasn't prepared to dish out the six hundred dollars. Everyone was clamoring to try one, but they had no demos available, which made it even more allusive.

Day 86: My thirty-ninth birthday is around the corner and I'm feeling the need to celebrate with friends, yet do something different and meaningful. So I came up with an idea to have a "Friendship Garden" planting party.

I sent out invitations today. The idea is to have each family that attends, plant seeds or seedlings in our backyard. We'll plant vegetables and flowers. Each "plot" will be marked with the type of seed planted on the front of the marker and who planted it will be written on the back. Then once the garden produces a harvest, my family will pick it and deliver the items to our friends that planted it as a way of continuing the friendship and doing something that's environmentally conscious. I can't wait!

Day 87: At the Community Recreation Center, across the hall from where Katelyn takes her singing lessons, there is a Game Room. It is packed with tweens fighting over the "Guitar Hero" game being played on a large flat-panel TV. The comfortable, worn couches are piled high with onlookers as the music blares out of the TV. Only two kids are playing pool, although there are several pool tables. Two young kids are coloring with crayons. The room is noisy as "guitarists" jam to Michael Jackson's *Thriller*. I inquire at the front desk about the Game Room. It is not open to the public. It's an after-school program for kids. Must be easier for the counselors to just switch on the TV, rather than engaging with the kids.

***

While looking over the Best Buy ad in the newspaper, Katelyn looked up at me and said, “Was there a time when they didn’t have TVs or cell phones?”

“Yes,” I said looking at her innocent face. “When I was your age, we didn’t have cell phones or the Internet.”

“What about movies? Did you have movies?”

“Yes,” I said.

“What about ballerinas? Was there a time when they didn’t have ballerinas?”

Day 89: Katelyn received some caterpillars in the mail today. I sent away for them as a birthday gift for her -- tried to go for something educational this year and steer clear of the commercialized presents. I can’t believe we actually had a *Hannah Montana*-themed party for her fifth birthday last year.

The caterpillars came is a see-through plastic jar and within three weeks are supposed to hatch into beautiful butterflies. Katelyn has enjoyed watching them wiggle and squirm.

***

We took Katelyn to look at her new school today. It’s closed, but we could play on the playground. She’s going from private kindergarten to public first grade. She'll not only have to navigate a larger class size, but also get accustomed to sharing lunchtime and recess with nearly 200 kids.

I'm nervous for her. She enjoys dangling from the bars and exploring the new bathroom. Looking at the garden and peering through the window of the cafeteria and media center (with rows of computers), I am suddenly tempted to pull out her iMac prior to school so she can brush up on her computer skills. I don't want her to be the only child not knowing how to use a mouse! This is Silicon Valley, after all.

Before we had unplugged, I had asked the school how we could meet some students her age over the summer so that they wouldn't all be strangers to Katelyn in the fall. The principal recommended that we join the school's Yahoo Group, which I did eagerly.

I sent a notice through cyberspace asking for summer play dates with first grade girls. A couple of mom replied and we did a few play dates before pulling the plug. Being unplugged makes me very uncomfortable because not only will Katelyn be making new friends, but it's a new social network for me too -- a very important one since our kids will be spending so much time together. Without Internet or e-mail, I could not unsubscribe from the group, so now, when moms send a message, I assume they'll get my auto response about how our family doesn't use technology…in Silicon Valley. I might as well have the plague. *What will these women think of me?* Also, the Yahoo Group is the main mode of communication for the school…functions, events, projects, and announcements. *How will I keep on top of things?* I think I'm more nervous than Katelyn about starting at a new school.

Day 90: Browsing the bookstore, a book caught my eye, *The Green Bible*. With a cover emblazoned with a green tree, it looked less intimidating than the black leather with gold embossing that it typical of the Bible.

I flipped through its thin pages. Every reference to the environment was typed in a green text. *Hmm.* Raised as a Christian, but married to an agnostic, my religious loyalties are on shaky ground. Maybe as I explore a life without technology, this would be a good time to use my downtime to delve into religion as well. I picked up a copy and added it to my reading list. At 1,251-pages, it would be quite an undertaking. I hoped the "green" twist would hold my interest among the out-of-date language.

***

Bree and wandered aimlessly down the street in the Haight district of San Francisco. With my scoop of Ben & Jerry's Chunky Monkey in hand, we took in the tie-dye clothing stores, smoke shops, tattoo parlors and vintage clothing stores amongst the dreadlocked shoppers and panhandling teenagers. One scruffy fellow caught my attention. A "homeless" man on a stool with a manual typewriter on his lap and a sign, "Custom Poetry." Not often you see a typewriter being put to use in public.

Month Four: Life Without a Net

"The internet has been a boon and a curse for teenagers."
--J. K. Rowling, author of the *Harry Potter* series

It has been three months, but seems like years since we've had our technology. But at the same time, it has flown by.

- TV: At this point I'd say that if it were only *my* decision, I wouldn't bring back the TV at all.
- E-mail: I miss being able to send group e-mail, but have been enjoying calling individuals instead.
- Computer: I love not being attached to any laptop. More calm.
- Internet: I miss getting info quickly and easily.
- Cell Phone: I don't miss my cell phone at all. I enjoy sitting down and taking the time to give my full attention to the conversation.
- Photo: Film is the too expensive, but digital is too time consuming.

Day 91: Bree is back from volunteering in Peru. She taught some kids art and English. She chose not to stay in touch by cell phone or e-mail while away, but rather just enjoyed the experience.

She's inspired me to volunteer with some time freed up by not being tied to the boob tube. I've decided to spend a few hours a week volunteering my time to a good cause. After weighing some options, (Humane Society, Bay Area Wilderness Training, Hidden Villa Farm) I decided to lend a hand working in an organic garden at a small farm nearby called Hidden Villa.

I've taken Katelyn for visits to this local farm a few times and she's always enjoyed romping through the flower patch, petting the goats, exploring the grounds, walking along the creek. I thought it would be nice to help take care of a place our family has enjoyed. I start tomorrow. Time to get my hands dirty.

***

I used to let Katelyn watch her favorite TV show while I took a shower, but without that as an option, lately she's been feeling like she needs to keep me company, which usually involves singing.

The other day I got out of the shower and was toweling off and she said to me, "Mommy, why do you have hair on your vagina?"

"Women just do," I said.

"Will I, when I grow up?" Katelyn asked.

"Yep," I replied.

It reminds me when she was about two and her and Jeff were out and he had to go pee -- Katelyn asked him, "Daddy, why does your vagina look so funny?" He's never taken her pee with him ever since.

Day 92: A traditional back-to-school outing today. I took Katelyn shopping for a new backpack and lunch box. At Target an aisle was overflowing with lunch boxes -- Spiderman, Scooby Doo, Jonas Brother on one side -- Hannah Montana, Little Pets, Princess, Barbie on the other. No surprise -- Katelyn chose a Hannah Montana backpack and Cinderella lunchbox. I tried to steer her towards the "cute pink one with flowers." She wanted nothing to do with it. While there, we browsed the kids' swimsuit section. Her one-piece Ariel suit, as cousin Marie thankfully pointed out, is "showing a little too much bun." It was riding up her bottom with her growth spurt. But she refused to give it up. After all, it was Ariel, her favorite princess.

Flipping through the picked over selection, I offered her rainbow, tied dyed, polka dot, striped bikinis and one pieces. She finally agreed upon a two-piece pink suit with silver trim and a skirt bottom. At the pool later that day, she commented, "I really wish we could've found one with a character on it." Marketing at its finest.

Day 93: Even if I never let Katelyn watch another *Hannah Montana* show or never bought her anymore *Hannah Mont*ana merchandise, there's no escaping the wrath of Disney. Don't get me wrong, the Disney channel has been my saving grace many times before unplugging. But *Hannah Montana* is everywhere -- on TV at my eye doctor, at Katelyn's pediatrician, at the car dealership. *Hannah Montana* clothes at Target and Kohl's. *Hannah Montana* toys at Target and Toys "R" Us. Her song on the radio. *Hannah Montana* cereal dishes, camera, vitamins, pens, notepads, stickers, shoes, waffles, candy, sunglasses. I think you could survive on *Hannah Montana* products alone.

***

The caterpillars have grown from babies to teenagers in just about a week. In the past, we've watched TV shows about the metamorphosis process, but Katelyn is thrilled and in awe seeing it happen in real life. She decided to bring them to summer school today for show and tell. I was so proud as she stood in front of a class of seven kids, held up the plastic jar and told her classmates how she's "growing butterflies." She then cautiously went around the circle letting the kids get a close up glimpse, but not touch "because we don't want to disturb the caterpillars." The life cycle is truly amazing.

***

I finally took a sewing class -- hoping to put my TV-free evenings to good use. Six students and two teachers crammed into the back of a fabric store storage room. I learned the parts of the machine, how to read a pattern and spent sixty-five dollars (fabric and lessons) to emerge from the class with a pair of pajama bottoms. Not exactly *Project Runway*, but it's a start. Meanwhile, Katelyn loves her singing lessons, but poor Jeff had to ditch his guitar class because he's been too busy with work.

***

I worked out with my trainer Brian today. He is whipping my butt into shape, literally. After he wrote out a new routine for me, we went back to his office to print it out from his computer, but his printer was on the blink.

"Can I just e-mail it to you?" asked Brian.

"Uh, I don't have e-mail," I said, embarrassed.

"You do know this is 2009, don't you?" Brain smirked. "Can I e-mail it to a neighbor?"

"Can you e-mail it to the front desk and have them print it?" I inquired.

"I'll figure it out and leave it at the front desk for you," Brian finally said.

***

Enjoying iced tea at the Roasting Company while Bailey took Katelyn for pizza and ice cream down the street, I overheard, "I'm so addicted to all these games on Facebook."

A friend replied, "I didn't know they had games. I know you can add friends."

"Oh my God… the games!"

I turned to see who was talking and smiled to see that it was two women in their sixties. Facebook was not just for kids.

After about two hours passed and I begin to worry, after all, I did just meet Bailey at the Apple Store. How well did I know her? My instincts told me not to worry, but the mommy in me said I better call to see where they were. I asked to use the phone in the coffee shop and was told, "There's a payphone across the street in the park." *That's convenient, I'd much rather pay to use a phone than beg to borrow one.* I took a walk around the park and didn't see one. I did see an enclosure that looked as if a phone might have hung there at one time. A woman and her teenage son were sitting on the bench.

"Do you know where the pay phone is?" I enquired.

"The what?" she asked perplexed.

"The payphone," I repeated, my face turning slightly pink with embarrassment.

"Kinkos?" she said.

"No, the pay phone," I said.

"Oh, those things are so obsolete," she said with a smirk. "I wasn't sure I heard you right. If you need to make a call, you can use my cell phone."

I agreed. *Was this breaking the rules?* No network coverage. The well-dressed lady dialed again for me. Busy. Redial. Voicemail. I left a message and profusely thanked the woman for her kind gesture. As I walked away, her phone rang.

"Oh, it's your friend calling," she waved to me.

"Hey Bailey, where are you guys?" I spoke quietly into the phone.

"At the pet store," she said. "Then we're stopping by the candy store."

"Oh, OK. Do you want me to pick you guys up?"

"No, we can walk. See you at the coffee shop."

Handing the phone back, it was nice to know that some people are still kind to strangers.

Day 94: I never thought I would, but I finally broke down and subscribed to *People* magazine. I just can't live without my smut. When I was in my late twenties, I was a high-technology career woman working in the big city (San Francisco) and, while commuting to work on the train one day, I overheard some young ladies yapping about the latest Hollywood gossip. I thought to myself, *that's so stupid that they give a shit about what celebrities are eating, who they're dating, what they're wearing, how much weight they've lost. There are much more important topics to discuss. How immature.* Now, in my late thirties, and being a mostly stay at home Mom, I think I just crave some easy entertainment and a change from mommy mode.

When I think about it, it's pretty pathetic actually. I used to waste my time watching reality shows and laughing about how screwed up the lives were of the stars, how they led such insignificant lives, yet what does that say about me? The only thing worse than living a life full of drama, is *watching* it. At least the reality show stars weren't sitting on the couch. They had their own show! I can't wait for my *People* magazine to arrive.

***

I had arranged for us to have a play date with Katelyn's friend Sandy. It was our first time to her house. Sandy's mom, Sage, and I sat on the couch as the girls performed a dance routine to one of their favorite *Hannah Montana* songs. I glanced over at the stereo and noticed some worn albums and a record player.

I said to Sage in amazement, "You have a record player?"

"Yeah, we bought it on Craigslist," she said. "I pulled out my old Michael Jackson album and listened to it in tribute to him the day he died."

I flipped through her stack of '70s records and pulled one out to show Katelyn.

"This is called a record or album," I said. "It's how we listened to music before CDs were around."

I placed it carefully on the record player. Katelyn watched curiously as the record spun around and around. I put the needle onto the record and heard Michael Jackson crackle through the speakers.

"My brother still has an 8 track that he listens to," Sage laughed.

Day 95: Lately a neighbor's dog has been barking incessantly. I'm not sure if it's something new or if I'm just noticing it because we don't have the constant noise of the TV to mask the barking. The irritating sound is permeating our peace. Katelyn said, "I bet the owner is on their computer and not giving the dog attention."

***

I started the volunteer job at Hidden Villa Farm today. I'm working for the horticulture manager, David. I arrived with my sun hat, water, gloves, and sunscreen -- prepared for hard labor. Joining me were two regular volunteers, two college interns, and one-day volunteers (a woman with her teenage son). The day started with David teaching us how to make gopher cages for trees we'd be planting, so the gophers wouldn't eat the roots. We wore thick gloves and bent chicken wire to form containers that we'd wrap around the trunk of the tree. Shaded by an Oak tree, we chatted about this and that. Next, six kids from the summer camp joined us. The "environmental heroes" had studied how trees help the environment. We paired up with the kids and planted native Yarrow and Hummingbird Sage.

David said, "Come back in a few years to check on 'your' trees and you'll see how much they've grown."

David has a gruff, scruffy beard, worn hands, and dusty clothes, but a gentle manner -- like a father figure. Next, we said goodbye to the kids and spent an hour pulling weeds. At first, I cautiously squatted, not wanting to get too dirty. Eventually, my thighs were burning -- no need for a gym. By the end of the day, I said 'F-it' and plunked myself down in the dirt. Getting grubby was part of the job.

The last hour was spent working in the educational garden. I swung the metal gate open. It had a hand-painted sign attached precariously that read, "Keep closed to keep deer out." Opening the gate was like stepping into Oz. The colors in the garden were so vivid, the sunflowers reached the sky and kissed the sun, there was an abundance of organic corn, carrots, lettuce, colorful signs describing what was growing, handmade scarecrows were fat with straw, apple trees, lavender, bees, the air full of wonderful smells. I felt alive and at peace simultaneously. Invigorated and grateful.

We scattered old seeds to see what would grow. As I left the farm, David said, "Thank you so much for volunteering your time." When, in fact, *I* was the one that was feeling thankful to be able to volunteer -- grateful to have the free time to help.

Day 96: We were about to embark on our long drive to Laguna Beach and no computer for Katelyn to watch movies on. We had to be creative. At 3:30 in the morning, Jeff poked me in the side as I lay snuggled up in bed under our down comforter.

"You up?" he whispered.

"No," I replied grumpily, pulling the covers tighter around my body.

"Let's go," he said.

I rubbed my eyes and tried to focus on the clock. "It's 3 a.m. You're crazy," I replied.

"I can't sleep," he said softly, while spooning me. "I might as well be driving. You and Katelyn can sleep in the car."

"You really want me to leave my cozy bed?"

"Yeah, let's go," Jeff said enthusiastically.

"Fine," I huffed. "Pack the car and I'll get dressed."

Katelyn and I slept for the first four hours of the drive. Arriving in Laguna Beach by 10:30 a.m., Jeff's family was shocked with our early appearance. Our visit to Laguna was wonderful a wonderful mixture of technology and nature -- playing Wii with the cousins, watching movies on TV and in the minivan, frolicking in the ocean, visiting rescued sea lions with the grandparents, grinding corn like Indians at Nix Nature Center, spotting tarantula wasps at Riley Wilderness Park, and spinning pottery at the Sawdust Festival. I hate to admit that it was nice to have a taste of technology again, but at the same time, I didn't mind going back to my now slower pace of life.

Day 101: Having just returned from Laguna Beach, we awoke to two butterflies that had emerged from their cocoons. Katelyn and I ran over to our young neighbor Felicity's house to tell her the news since she watched they while we were away.

She said, "I guess they wanted to wait for you to get home."

By the end of the day, all five had hatched. Katelyn was thrilled. We read the instructions and fed them orange wedges on carnations sprayed with sugar water. What an amazing gift to see such a transformation.

Day 103: At Cin-Cin with Jeff, I enjoyed wine tasting as the chatty bar buzzed with activity. The waiter brought us a watermelon and feta salad with an olive sampler -- the perfect match for our flight of white wines. After a few sips and nibbles, Jeff said to me, "I think I did a bad thing."

"What?" I looked at him -- his eyes turned down, looking at the wooden bar.

"My friend from work sent the GAP friends and family thirty percent discount to my work e-mail and I printed it out," he bit his lip, awaiting my response.

"Oh, why did you have to tell me that?" I said with a grimace. "Of course, you're breaking our 'unplugged' rules…but I've bee waiting for that coupon for months!"

"I better just rip it up," Jeff taunted.

"No!!! The Banana Republic outlet is having their forty percent off sale right now. Please, can we keep it?" I begged.

"Nope. It would be cheating."

"Can't you just ask her print it out for you?"

"No, that would be weird," he said.

"Yes. Being unplugged forces you to face situations that are uncomfortable. Come on. Man up. Take one for the team," I said, lightly slugging him on the shoulder.

"Nope. It's done. No coupon."

I pouted, but was proud of him for trying to honor what we were doing. *Damn it.*

***

Sitting on a bench in the sun outside of the Los Gatos Coffee Roasting Company, I was indulging in a good book when a gentleman sat down next to me. His cell phone rang. After a few seconds, I heard him say, “Are you there? Can you hear me?” No response. He hung up. A few seconds later his phone rang again.

I was wrapped up in my book and not concerned with his conversation. Once he hung up, he turned to me and said, “I wasn’t talking too loud was I? I’m trying to be conscientious of others while I’m on my cell phone.” *What a novel thought.*

While living in Australia in 2000, it was considered utterly rude to even bring a cell phone into a restaurant, more or less talk on it while at the table. Apparently, that attitude has not caught on in the U.S. *Would it ever?* Even when I did use my cell phone, I would try not to answer it when I was with someone in person. Face-to-face communication takes precedence.

Day 104: Driving down the street, I burped after swigging some Diet Coke too quickly. Katelyn piped up from her car seat, “Remember how

Daddy could make his iPhone look like he's drinking a beer, then it would burp." Quality apps that we can't live without.

Ironically, in 2008, CNN Money called "iBeer" a "must-have" app and it made Apple's "Top Paid Apps (Overall)" and "Top Paid Entertainment" lists. Luckily, there's also the "Drunk Sniper" app that simulates peeing in a toilet. The catch, "the more virtual drinks you have -- the more difficult it is to keep the spray within the toilet plus the toilet starts moving and blurring with every additional level." Technology at its finest.

***

I feel like a honeymooner. Jeff and I have had sex for five days in a row. That hasn't happened in years (since before becoming parents)! The first couple of nights, it was like, "Yeah, OK, it's good, but are we done yet?" By the third night, I was initiating it. On the fourth night, I got daring and had a date with Jeff and left my underwear at home! Flashing him in the car, he was beside himself. Needless to say, the date didn't last long because we couldn't wait to get home. By the fifth night, it went from being a marital obligation to fun, flirty and casual. Somehow, as a parent, I'd come to think of sex as something serious and a lot of work. Now, I'm seeing it as a playful way to end the day. It doesn't' have to be hot and heavy. We can stop, chat, laugh, while having sex. Nice. For me, I'm finding sex is like chocolate -- the more I have it, the more I want it.

***

A mommy-and-me trip to Fresno to visit my parents meant hours for Katelyn to entertain herself in the car. She listened to two books on tape, colored in her Barbie coloring book, played her "piano," read books, sung *Hannah Montana* songs. I have to admit, giving up TV after your child knows how to read is very helpful. *What would it be like for a teenager to give up technology?* Unbearable, I imagine.

At Mom and Dad's, they wanted to take me birthday shopping -- a trip to Lowe's to pick out some gardening tools -- a shovel, hoe, knee pads and rake, and I a mini shovel and hoe for Katelyn. Dad and I had chatted the night before about gardening.

"I loved our large veggie garden when I was a kid," I told him. "How did you learn about gardening?"

"When I was a child, my grandmother had a garden and she would send my brothers and I around the neighborhood with a wagon to pick up fallen leaves for mulch," he said, telling me something I had never known about his childhood.

"I can remember picking pumpkins, zucchini, tomatoes, carrots from our patch in Eureka," I said. "It was so much fun to watch them grown. I want to share that experience with Katelyn."

I guess that makes me a fourth generation gardener. Gardening seems to be a common bond that brings people together. Now that I'm aware of it, I'm finding that a lot of my friends have gardens and now we have that experience to share.

At my parents' house, Katelyn was glued to their large screen TV, watching *Camp Rock*. While unpacking our luggage In my old bedroom, I saw my record player covered in dust.

I yelled to the living room, "Dad, does this record player still work?"

He came in and said, "I don't know. It hasn't been used since you moved out -- what…20 years ago?"

I wiped it clean with a cloth, plugged it in and gently placed the needle on the spinning album. It worked. The static of the record was nostalgic. Next to the player, I flipped through the record collection that had belonged to my parents -- Beach Boys, Shaft, Rolling Stones, Woodstock – along with my Junior High collection of Michael Jackson, Shawn Cassidy, and Depeche Mode.

"Can I take this home with me? I think Katelyn would get a kick out of it."

"Of course, it's yours," Dad shrugged.

It's funny how all these years that I've been back to visit my parents and never once had an interest in the record player. Never really noticed it. Thought of it as junk, really. Now, suddenly it was a prized possession to be treasured. My outlook on life was changing.

While packing up, I remembered that my high school yearbooks used to be on the shelf near the record player, but I didn't see them there.

"Dad, do you still have my yearbooks? It would be fun to show Jeff," I said to him while he was helping me gather my belongings to load into the car. He dug them out of a top shelf in the closet.

"1988?" he said looking at a weathered book. "This can't be yours. You can't be that old."

I smiled knowing exactly how he felt. Katelyn will be in high school in the blink of an eye.

"Oh, and if you want, I'll finally take my trunk full of junk off your hands. Is it still in the garage?" I asked him as I flipped through the yearbook, gawking at the old photos.

I don't know what age I was when I started keeping mementos locked in a black trunk, but for the past ten years, since moving back to California from Hawaii, my parents have been trying to give my junk back to me. Until recently, I never lived in a house with a garage. Now that I had the space, it might be fun to open the time capsule.

We headed to the garage, my dad looked around trying to remember where it was stored -- under a pile of boxes, of course. He unstacked the boxes and pulled the rusted, worn trunk out of the garage and heaved it into the trunk of my car.

"I tried to open it once, to make sure there wasn't something in there that would spoil, but I couldn't get the lock off," he admitted.

The lock was on their for a reason…to keep my parents out! As a parent, he must have been dying to know what secrets I was keeping locked inside.

Day 105: "Can we use our gardening tools today, Mama?" Katelyn asked, poised to get her hands in the soil.

"Sure, we need to get the soil ready for planting," I said with a smile.

"I'd rather have Dora gloves, than ladybugs," she stated, wiggling her fingers in her new gloves.

We raked and hoed the large patch of soil that would soon be home to our "Friendship Garden." I was amazed at her strength, endurance and dedication. She was a hard worker. No whining. She was joyfully getting dirty and sweaty.

"Is gardening exercise? It's a lot of hard work," she said wiping the sweat from her forehead.

As she used her pink hoe to loosen the dirt she said, "Mama, I like gardening because then I can see what you do for your (volunteer) job."

I had no idea that volunteering my time would inevitably lead to bonding with my daughter. What a wonderful gift. Katelyn has always been an affectionate child, but has become more so in the past couple of months. When she's serious, she calls me Mommy. When she's lovey, I'm Mama. When she was a baby, I was Mimi.

Lately, when we spend quality time together she'll say, "Mama, I love you one hundred percent."

After gardening she said, "Mama, I love you nine hundred percent," and wrapped her arms tightly around my waist.

The interesting (and sad) change is that she has always been Daddy's girl, but since unplugging she's shifted to preferring me. A combination of me giving her more attention (paying less attention to e-mail and TV) and Jeff working more evenings out of the home since we don't have Internet access. I feel awful. I'm happier than I have been in years and Jeff is miserable.

Day 107: Katelyn and Jeff are both asleep. I'm propped up in bed with the lights down low. Luckily, Jeff is a sound sleeper. Once he's out, I could throw a disco party in our bedroom and he wouldn't rouse. On my nightstand, my yearbooks are piled high -- 1985-1988. The most pivotal time in my life was 1986. I was a sophomore in high school, when my Dad was offered a job at Fresno State as an electrical engineer, 457 miles away from the small town of Eureka where I had spent my childhood growing up.

The job came with a much-needed raise that would financially change things for our family. Little did my parents know, the move would devastate me. Flipping through the pages of my freshman year at Eureka High, I landed on my class photo, which showed me with my naturally dark brown hair in a feathered '80s style hairdo.

By the time my sophomore yearbook arrived, I had already moved to Fresno, but came back to Eureka to pick it up and see my friends at the yearbook signing party. As I lay quietly in bed reading the inscriptions, I was instantly transported back in time. Scribbled on the back page, several friends had written, "EHS isn't the same without you!" My chest grew heavy and my eyes welled up. Even 22 years later, the pain came rushing back. Saying goodbye to friends I had known since kindergarten had been unbearable. As I lie in bed, contemplating my high school friendships, it seemed odd to me that they were so deep and meaningful at the time, yet eventually they faded into nothingness. I couldn't help wonder if I would've kept in touch with those dear friends had I grown up in the Internet era. *If e-mail would have been an option at that time, would I have been able to maintain those friendships?* My thoughts fast-forwarded to my current state of living unplugged and I cringed. I longed to look up old buddies on Facebook and reconnect.

How could I have let them go? Facebook suddenly had more meaning. I needed to find those I had lost.

When I had convinced Jeff to join Facebook, we had a contest to see who could get the most "friends" – 5,000 is the limit Facebook would allow, although some people don't think that is enough. With only 360 friends, I felt unaccomplished. It's great to locate people from your past, but isn't it a better use of time to just call friends who currently mean the most to you, rather than wasting time posting to everyone you've ever met that you had a hard-boiled egg for breakfast? Quality versus quantity.

Day 108: Ironically, a few days before unplugging, my friend Heather sent me an e-mail saying that she was starting a nature-based environmental education program for kids. She named it "The Caterpillar Kids Club." Kids would attend weekly classes to learn about things such as planting, hiking, recycling. Having just read Richard Louv's book *Last Child in the Woods*, I saw the importance of what she was doing. I had recently mailed her a card just to keep in touch and she called me immediately.

"Sharael, thank you so much for the handwritten note. I don't get much of those anymore," she said, delighted.

She invited Katelyn to attend one of her classes and we accepted. We arrived at Gamble Gardens in Palo Alto at 10 a.m. and the Caterpillar Kids were scattered on a blue tarp shaded by a large Oak tree. Heather started the class by strumming a native-themed song on her guitar accompanied by the half dozen toddlers and moms. Next was garden bingo. She handed out a sheet with photos of objects in the garden and stickers to mark when they were found. Katelyn loved it. It kept the kids focused and interested. We searched for sunflowers, pumpkins, a fountain, and a sundial.

As we walked, Heather informed the kids about various aspects of the garden. We felt a soft lambs ear plant, saw a spider eating a bug, learned how to tell time on a sundial. Next, we sat around a picnic table and each child scooped soil into a disposable cup, then popped in wildflower seeds, added more soil, sprinkled it with water and was encouraged to take it home and tend to it. The class ended with Heather reading nature books and singing more songs. A part of me wished Katelyn was younger so that we could attend these classes weekly, but with summer coming to an end she'd be busy with schoolwork soon.

I left feeling even more appreciation for nature and grateful that Katelyn has had the opportunity to have been exposed to it, unlike some children that prefer video games to hiking. My mind began to ponder how I could also help educate children about the importance of nature.

***

The day of Katelyn's singing recital had arrived. She was frightfully nervous. There was so much pressure because all the parents would be watching. I tried to reassure her and calm her fears.

"It's natural to be nervous, but you'll be fine," I told her.

At the recital, the kids practiced with the teacher a few more times before the parents were allowed to enter the room. Katelyn stood stiffly, her eyes pointed down at the floor. Halfway through the first song, Katelyn was singing softly with tight lips and her arms folded across her body. As I watched her painfully put on a brave face, my heart sunk. Soon her face became flushed and her eyes filled with tears. Jeff held out his arms for her to come to him. He swooped her up mid-song and whisked her out into the hallway where she began sobbing, "I just didn't like all those people staring at me."

Rubbing her back, he gently reassured her that a lot of kids feel the same way but told her how proud he was that she was brave enough to even try. When we had arrived, I was once again in agony over not having our video camera to record her first singing recital, but after her melt down, I felt grateful that she didn't have the added pressure to perform in front of the camera.

While we had been waiting for the recital to begin, another couple had arrived to watch their daughter -- the mom was dressed in business attire and carried a laptop. She sat on the hallway bench and opened it and began tapping away. Her husband held a digital camera and was in search of an outlet to plug it in to charge it. While it was charging, he toggled on his iPhone. Watching them was like looking in the past. Just a few short months ago, that was us! As we waited, Jeff and I had no cell phones, e-mail or laptops to distract us. Instead, we had each other's undivided attention and relished it.

Day 109: Another day on the farm. The rain soaked my windshield as I drove down the highway wondering if I should turn around. Upon arriving, I spotted David at our usual meeting spot.

"Need a hand?" I called

"Hey Sharael, great to see you," he greeted me.

The other volunteers arrived and we were assigned our daily duties -- help some kids plant trees then either do some gardening or help bake pies for the upcoming fundraising dinner. I jumped at the chance to do some baking. A young woman named Alice joined me. David took us to the historic farmhouse where the interns lived, so that we could use their kitchen while they were out tending to the farm.

Our task was to bake two peach-blackberry pies, two strawberry-rhubarb pies and some bread. The historic house was home to 10 interns that lived there for one year. Entering the kitchen felt like stepping back in time or like being in a farmhouse in Nebraska, yet here we were in Silicon Valley in 2009. The interns lived off the farm -- literally. The fridge was filled with raw goat's milk (as noted by the handwritten label), and fresh eggs from the chickens, fresh veggies -- straight from the garden -- were on the table. There was a large compost basket under the sink, leftover canning jars used as drinking glasses, homemade pickles, and recycling bins under the counter.

A large chalkboard designated who was responsible for collecting eggs, emptying the compost, washing the cloth napkins, cooking dinner, etc. As Alice and I read from our recipe David had printed off the Internet, we were in awe that people actually lived this way. No Ziploc bags. No processed food. No paper towels. A simple, uncomplicated life.

As I sliced the rhubarb from the garden, Ben came home for lunch. He concocted a stir-fry from an assortment of fresh veggies. I was impressed with his resourcefulness. An onion, a carrot and some potatoes -- actually looked healthy and tasty.

I said jokingly, "So, do you kill the chickens?"

"Yeah, sometimes," he admitted. "That's part of my job. I don't mind. I'd rather know what I'm eating -- how it was raised."

"So, do you guys get a stipend in addition to room and board?"

"Yeah, but it's not much," he replied. "The main thing is we get free food."

I left with a new appreciation for organic farming.

Day 111: We were fortunate enough to be able to attend Jeff's coworker's wedding at Kirkwood in Lake Tahoe. The ceremony took place as the sun set behind the mountain. Guests were cautioned not to wear high heels since it took place in a field of grass. The ceremony was the most untraditional, but touching wedding I've eve been to. No bridesmaids or groomsmen -- instead family and close friends presided the bride, the women carrying bouquets of wildflowers -- no matching outfits. The bride sauntered down the aisle in a beautiful, very traditional white silk gown with a veil covering her smile. The crowd was wedding casual -- suits with flip-flops. Jeff looked so handsome. It had been a while since I'd seen him decked out. For me, a silk beige dress that was an anniversary gift from Jeff with a black pearl necklace that he bought me on our honeymoon in Tahiti.

Friends and family took turns informing the crowd of how the couple had met and what they wished for their future. Vows were said, rings were exchanged -- sealed with a kiss -- Katelyn giggled. One well-spoken man had been the bride's host father when she studied abroad in Germany. He begged the question, "We are family. What is the definition of family? Look it up on Google."

At the wedding reception, the bride and groom assigned seats at "themed" tables. I laughed when I saw that we were seated at the "We Love Broadband" table -- how ironic. Four months ago, that would have been a perfect description of us and our life in Silicon Valley. *My, how things have changed.* The bottom line -- we do love broadband, we just don't use it anymore. As the bride and groom did their first dance, I snapped photos with my disposable camera. I had planned on bringing my 35 mm, but upon arriving at the resort I noticed the batteries were dead. At the resort's general store, they didn't sell "specialty" batteries, but luckily they did sell disposable cameras. Prior to unplugging, I would have been very disappointed to not be able to capture the wedding with the creative flair my real camera would have allowed, but I've become much more relaxed about things not going my way and have learned to become flexible and go with the flow. Normally, I would've made Jeff drive me one hour to the other side of Tahoe for batteries.

The weekend was beautiful and relaxing -- a gondola ride to the top of the mountain, hiking through the woods, relaxing by the lake, casual reading. Times like these were worth slowing down for and remembering.

Day 113: Katelyn has a cold -- always a challenge without a TV to occupy a lethargic child. Katelyn relaxed in the shade of our backyard in the zero gravity chair. I gave her some of her favorite books. Meanwhile, I indulged in some gardening -- a butterfly garden to be home to our newly raised butterflies. I was surprisingly thankful for not having a TV on a sick day.

"Mama, thank you for saying yes to having me," Katelyn said, her eyes sparkling.

I melted. "I didn't have to say yes. I always wanted you," I smiled.

Katelyn was becoming extremely affectionate and surprisingly independent. Now she's calling me "Mama" 24/7. And she's again sharing the love with Jeff.

We went to the Human Society today to fill out their online application for foster kittens. I thought it would be a nice distraction and comfort. Now we wait for kittens in need.

Day 114: "Katelyn, I have to tell you something. The butterflies are getting near the end of their lives and it would be nice if they could experience flying free outdoors before they die, so it's time to set them free," I told her.

With a scrunchy face and clenched fists on waist, she belted, "No! I will miss them. You can't let them go."

Jeff chimed in, "It's the right thing to do. They shouldn't spend their whole lives trapped in a small space."

"We can let them go in our new butterfly garden," I said. "Maybe they'll stick around."

Hesitantly, Katelyn said, "OK. Can I see if they'll sit on my finger?"

"Of course," I smiled.

We carried them to the backyard and unzipped their house. They were reluctant to fly away. As each one flew out, we yelled, "Bye, bye, butterfly." Katelyn chased them as far as she could.

Day 115: 5 a.m. my alarm goes off -- *beep, beep, beep*. I drowsily press the off button and roll over, snuggling into the down comforter. I reach for Jeff, he's not there.

"Love?" I loudly whisper.

He tiptoes into the bedroom.

"What are you doing up?" I ask.

"Working," he says quietly, so as not to wake Katelyn.

"How long have you been awake?"

"About a half-hour."

"You're working at 4 a.m.?"

"I have a PowerPoint presentation to finish for a meeting this morning," he says.

He walks back to the home office. I lay in bed staring at the clock -- 5:16. I fight with myself.

Angel: *Get your butt out of bed and go to the gym.*

Demon: *But it will be so boring on the treadmill without TV or music.*

Angel: *But you're getting fat! You've already put on five pounds in the past two months.*

Demon: *Trying to read lips on the large screen TVs at the gym is torture. Definitely not worth leaving my cozy bed.*

I give up. I just can't do the indoor cardio workouts anymore. I tried for at least a month. Enough is enough. I need to return to nature. Brian is not going to be happy with my lack of adherence. So much for my six-pack abs.

***

I read in a Bay Area parent magazine that there is a San Jose Quilt and Textile Museum. With a new interest in sewing, I thought it might be inspirational, so I made a field trip out of it. Katelyn, who has become completely in awe of my sewing ability (which is very basic and sloppy), was surprisingly interested in the museum. She has recently been begging to stay up late to watch me cut out patterns or sew.

In the entrance there was a dress made from a quilt. It was pink with large flowers. Katelyn loved it.

"Mama, can you make me a dress like this?" she beamed.

I looked closely at the design, "I can try."

The actual quilt exhibit included quilts displayed on the walls of the museum. I've never really paid much attention to the intricacies in the design, but was impressed and gained a new respect for quilters. Maybe I'd try my hand at a simple patchwork design.

Down the street, we visited the San Jose Museum of Art. As we wandered through a mobile exhibit of sculptures and more paintings, a few exhibits had pre-loaded iPods and headphones with QuickTime movies about the artist.

We came across a rack with bags that begged to be borrowed. Inside were sketch pencils and erasers, pastel crayons, and a notepad -- for guests to borrow. In an adjacent bin were large clipboards with drawing paper. I grabbed a bag and encouraged Katelyn to make a sketch of one of the paintings. She chose one of a piece of chocolate cake. We sat on a black leather bench in the middle of the room and she studied the painting and tried to replicate it as best as she could. She'd get up and go take a closer look, then come back to the bench to work on her masterpiece. I shared the clipboard to draw the ice cream sundae painting. When I was finished Katelyn marveled at my artistic ability.

"Wow, Mama. I didn't realize you could draw so well," she said.

It didn't take much to impress her. I reveled in her curiosity about a side of me she'd never seen.

"Can we go to the room with the animal cookies? I want to draw those," she said, standing eagerly.

In a large room were sculptures replicating large animal cookies with pink and white frosting and rainbow sprinkles. There was no bench, so we sat crisscross applesauce in the middle of the floor. Katelyn eyed the sculpture from every angle. As she worked on her drawing, her face became intense on her work. She lay down on her stomach, bent legs in the air. Before I knew it, the two of us were sprawled out on the floor -- our art supplies strewn across the hardwood as museumgoers strolled past. Probably a bit inappropriate, but we were having a moment -- making a memory. I vowed to bring our own art supplies to museums in the future. It made the trip go from "this is boring just walking around looking at paintings" to "look what I made. It looks just like the real one." A truly interactive experience. Maybe we should do that with more things -- make sketches at the park, the market, the zoo.

Day 116: I took Katelyn to Goodwill to show her where our donations ended up after our garage sale and explained how the money raised by selling our used items would help underprivileged people get education and skills training to find jobs.

While there, I spotted racks of records! Sixty cents for John Denver, Saturday Night Fever, Mary Poppins, Garfunkel. *Score!*

Day 118: Turning 39 wasn't monumental for me. Normally, my birthday is a family dinner or a girls' spa day, but this year, I was excited about planting our "Friendship Garden."

Jeff and I had gotten up at 8 a.m., pre-digging the holes for planting. The ground was very hard. Quite an undertaking. Katelyn lent a hand. We put "Sweet Pumpkin," our scarecrow, in the middle of the garden and a hand-painted sign that read "Friendship Garden."

I gathered my guests around and said, "This is a Friendship Garden because my friends are helping me plant it. I'll look out into the garden every day and be reminded of what wonderful friends I have. Originally, I had wanted to plant vegetables and when they were harvested, I was going to bring them to your homes -- to thank you for the planting and for your friendship. But since we're in the off-season, you'll be getting Basil and Rosemary instead!"

Jeff gave the group a demo on how to plant. The kids really enjoyed digging in the dirt, shoveling, planting, watering. With the temperature in the high 90s, I used the hose mister to cool the kids down after their hard work in the garden.

Next we dined on croissant sandwiches and farmer's market grapes with lemonade for lunch. Afterwards, the kids hand-painted marker signs for the garden so we could remember what we planted. Once the colorful signs were placed in the garden, it really came to life. A work of art! Jeff and Bree brought out a decadent chocolate cake and everyone sang Happy Birthday. I felt so blessed to have such dear group of friends.

Day 119: David's girlfriend had invited me to attend service at the Conscious Living Center in Mountain View. I decided to take her up on it.

Upon arriving, she greeted me and said, "Oh, I'm so glad you decided to come. We went to the early service, but let me park and I'll walk you in and introduce you."

She gave me a hug and walked me inside. The Center was located in a business park, typical Silicon Valley style. The entryway had literature and greeters. Alice introduced me to the Reverend and some others who informed me that their center was a place to "test your wings, let your light shine, and learn to live the life of your dreams with joy-filled confidence, knowing you are not alone."

A fountain bubbled in the background. Candles glowed. It felt more like a conference room than church -- no stain glassed windows. A side room was dark and quiet -- used for meditation. There was a kitchen, offices and childcare. Ironically, the sermon touched upon gardening.

Month Five: Lap-topless

"Computers are useless. They can only give you answers."
--Pablo Picasso

Day 122: "I don't think I'll be able to make it to the circus tonight…I just have too much work to do," Jeff said over the phone.

"Oh no, Katelyn will be so disappointed," I said. "She was really looking forward to you joining us."

"Fine," he huffed. "I'll go."

"No, I understand. I'm not trying to make you feel bad," I reassured him.

"It's just so hard to fit it all in. I'm torn."

"I just don't want her to look back and feel like Daddy wasn't there when we did special outings. I mean, this *is* her first circus."

"What time does it end?" he questioned.

"I don't know, maybe it's an hour. We should be home by nine."

Jeff got home from work and we quickly piled into the car to headed off for the circus.

"Can you drive so I can get some work done on the way there?" he asked, sitting in the passenger seat with his laptop on his lap.

"It's only ten minutes away and it's not a safe place to leave your laptop in the car," I warned him.

He flung open the car door and stomped back to the house to return his laptop.

"Why can't you and Daddy just get along?" Katelyn pouted from the backseat.

"We're not fighting," I told her. "It's just hard for Daddy because he wants to be with you, but his boss needs him to do some work."

"It sounds like his boss is a bully," Katelyn said, trying to stick up for her Daddy.

As the tigers roared and the trapezists flung themselves through the air, Katelyn watched in awe as she scooped cherry shaved ice into her mouth from her elephant souvenir cup. She was mesmerized by the elegantly dressed ladies riding high on enormous elephants and clowns that flipped themselves into a vat of whipped cream.

Jeff looked tense. I could tell he felt it was a waste of his time. At eight-thirty, they announced an intermission. I suggested we leave because it was "after bedtime," but was met with huge resistance from Katelyn. I was torn between wanting her to stay and being able to let Jeff get home.

"It's my first circus. We have to stay to see all of it," Katelyn begged

I suggested that Jeff take a taxi home, but he was reluctant to dish out the thirty dollars. So, we endured a few more acts after intermission and headed out before the mass exit. With Katelyn crashed out in her car seat, Jeff opened up about work pressures.

"I just want to look for a new job, but how can I do that without Internet access or e-mail? For all I know, old colleagues or headhunters could be e-mailing me with opportunities but I don't know about it because I can't check my e-mail," he complained.

"Well, we make the rules and we've always said we can use technology for work, because we have to make a living," I reassured him.

"Well, for now, I just need to get my work done for tomorrow," he replied, looking exhausted.

Jeff was up until 1 a.m. that night, then awoke at 5 a.m. to head into the office. I felt awful for suggesting he join us at the circus.

Day 123: With Katelyn off of school, I decided to bring her with me to volunteer at Hidden Villa. David greeted us warmly and apologized that he needed to attend a fundraiser meeting. With a list of projects in hand, Briden, Katelyn and I piled into the golf cart and headed over to work in the garden. Armed with ladybug-adorned gloves, a large brimmed hat and skin slathered with sunscreen, Katelyn was ready for garden duty.

We trimmed Yarrow and pulled wild clover. She had no problem lending a hand. After an hour, she said she was hungry and I suggested that she pick an apple from the tree for a snack, lifting her up to reach the apple of her choice. I hacked away at a Jerusalem sage as Katelyn munched on her nature's treat. Briden had gone to the tool shed to get more robust clippers.

Suddenly Katelyn yelled, "Mommy, my loose tooth…it's coming out."

I bent down to examine her tooth and could see blood at the gums. It was time. Dropping my shears, I grabbed her hand and led her to the bathroom nearby. Using some paper towels, I carefully grabbed the tooth and gently tugged…out it came. I dabbed Katelyn's bloody gums.

"Oh, it feels weird," she said running her tongue along her bare gum.

We returned to the garden and Katelyn gleamed while showing Briden her missing tooth. Not many kids can say they lost a tooth while eating an apple picked fresh from a tree during volunteering at a farm.

Day 124: On my answering machine was a lengthy message from my mother-in-law asking for advice on how to get her printer to work. It was on the blink after she changed her ink cartridges. As I sat listening to her message, I remembered the endless times I had a similar situation and spent hours trying to correct the problem. I felt a sense of relief to be living without technology and the headaches and frustrations it can present. By the time I was able to return her call, she had already spoken to Apple and was able to correct the problem.

"I'm so proud of myself," she said exuberantly. "I followed their directions and downloaded and installed nine programs, then unplugged and replugged everything in a specific order, and now my printer works. They are so smart and patient with someone as technology illiterate as me."

"You know more than you think you do, but I'm so proud of you for getting the help you needed and doing it," I congratulated her. "Good job. By fixing problems, you're learning even more about your computer."

Day 125: "What do you guys want to do with our afternoon?" Jeff asked.

"Let's go to the dump!" I said enthusiastically.

In unison, Jeff and Katelyn said, "No!"

"It's really interesting. I promise," I told them.

"It'll be too smelly," said Katelyn.

"But I have some things that the garbage man won't take -- some fluorescent light bulbs and old recycling bins. It's really cool. Don't you want to see where the garbage man takes our trash?"

“No, not really,” Katelyn said.

“I really don’t feel like spending my day off at the dump,” Jeff added.

“Fine. How about going to the Japanese Tea Garden in Golden Gate Park?” I suggested.

“Mama, why does everything these days have to be about gardening?” Katelyn whined.

“It doesn’t,” I insisted. “I just thought it would be fun. OK, how about fishing? We haven’t tried out your new Barbie fishing pole you got for your birthday.”

“Yes!” said Katelyn, jumping up and down enthusiastically.

“Where do we buy bait?” I grabbed the Yellow Pages and looked up “fishing supplies.” Who needs the Internet when you have the phone book?

I called a store, “Do you sell bait?”

“Yes, bobbers, weights, night crawlers, we got it all,” said the man on the other line.

“Do we need a license?” I inquired.

“Yeah, we sell those too. Ten dollars a day,” he replied.

After getting our license. We headed to nearby Loch Lomond County Park, where we bought a bobber, some eggs, a hook, spinner, and weight. We sat on the bank of the lake and cast our line into the water. No bite. The ranger, who sold us the gear, didn’t believe in fishing. I don't either, but I think *nearly* everything is worth experiencing once.

We decided to rent a motorboat to take out on the lake. I lounged across the seat with a life jacket as my pillow as Jeff taught Katelyn how to steer the boat. It was her first time fishing and being on a motorboat. I lay in the sun with my hat shielding my eyes as Jeff and Katelyn fished. Songs from elementary school came to me and I started singing out loud.

"Feeling groovy...la,la,la,la,la,la,la,la...feeling groovy," I burst out, remembering my sixth grade teacher teaching us the Simon and Garfunkel song as he strummed his guitar.

Katelyn thought it was so funny to see Mommy singing in a boat and wanted to be "just like Mommy," so she laid next to me and starting singing too (*Hannah Montana* songs, of course). A wonderful technology-less day, which we documented with our disposable camera.

Day 126: Katelyn was taking her nightly shower, which gave Jeff and I a small window in which to have a private conversation.

"I'm not doing well," I admitted to Jeff.

"What's up?" he said, scooting closer to me on the couch.

"This unplugged lifestyle is really frustrating. It's getting me down. I can't look up directions online, use my cell phone or make calls while I'm out. I'm losing touch with my friends. It's really getting to me," I said, resting my head on his warm chest.

"Yeah, tell me about it. It's tough for me too."

"I just don't know how to get out of this funk," I admitted.

"Hang in there," he said, stroking my hair. "There have been a lot of positives that have come out of this experience too. I know it's not easy, but it's not forever."

After Katelyn finished her shower, I wrapped her in a towel and sang the "Barney Song" (a tradition she hadn't outgrown). She got on her jammies and came into the office. I was sewing. Jeff was working. She sat on my lap, and with an intuitiveness that borders on being scary, she looked at me and said, "Mama, I just love how you are so good at so many things…sewing, gardening, writing." Maybe being unplugged wasn't so bad.

Day 127: First day of first grade at a public school. A new world for Katelyn -- and for us. Throngs of parents and kids marched to school -- backpacks full of new lunchboxes and supplies. Several parents were video taping the experience. As we left the house that morning, I hadn't even thought about video taping -- and Jeff had to remind me to grab the camera to take photos. We fought for a parking spot. The playground was filled with screaming kids and ogling parents.

The bell rang and we walked Katelyn to class and helped her find her desk. She was greeted with a big hug from her friend Sandy. We knew she was going to be in the class, thanks to a class roster they posted on campus a few days earlier. Thank God they hadn't e-mailed it. The kids were given nametags. Each desk had markers, a worksheet and a box with scissors. With four desks arranged together, Katelyn got to know her desk mates, Kara (who she met on the playground previously), Megan (who had just moved a week ago from Boulder), and Justin (someone she hadn't met before). The classroom buzzed with activity as parents kissed their kids goodbye and hoped for the best. Upon leaving the school, Jeff broke down in tears while driving home.

"I'm just so proud of her. She did so well. No hesitation," he said, wiping his face.

"I know," I consoled him. "I'm amazed at how brave she was. It can be so scary and intimidating starting a new school."

Day 128: Reading through the packet of info that Katelyn's teacher sent home, I realized that there was a school handbook on their website, but "a printed copy could be requested." It would be useful to have, but a bit embarrassing to ask for it. Also included in the packet was the school hot lunch menu. Katelyn inquired about it, so I called to get more info since I couldn't access the website. I was informed that lunch accounts could be set up online where money can be added to the account.

"Can we just pay cash?" I asked the receptionist.

"Sure, just put it in a baggy and have your child hand it to the cashier," she informed me.

Unfortunately, we were only given a September lunch menu, but it was August. After a couple of phone calls, I still wasn't able to find out the August menu, so Katelyn decided to wait until September to give hot lunch a try.

Day 130: At the school, moms are gathered around a tree in front, waiting for the bell to ring so we could swoop up our little ones after a long day of being in class. My friend Sage walked up.

"So, I signed up to volunteer in the classroom on Fridays," she said.

"How did you do that? I'd love to volunteer," I said.

"They sent a note to the Yahoo group. Oh, but of course you didn't get it since you're not online," she replied.

My survival skills kick in and I pry more info out of Sage.

"Is there another way to sign up?"

"I'm not sure," she said.

"I thought maybe next week at back to school night they might have sign-ups."

"You really need to get back online," she urged. "I understand not having the TV, but you need the Internet."

"I'm just not ready. It's too easy to get sucked back in. Two things I can't have in my house -- chocolate and Internet access. I'll over indulge in both. Do you know the name of the woman who sent the message?"

Sage said laughing, "I can look it up for you. I'll try to get her phone number too."

***

My father-in-law called. They are planning their first trip as retirees. Originally they wanted to drive from Vancouver to Laguna, then it got cut to Vancouver to San Francisco, and now it's down to Oregon to Eureka.

"Hey Share, we were wondering if you could recommend places to stay in Oregon since you've been there," my father-in-law asked. "The Auto Club told us we should fly into to Astoria and take Highway 101 south, down the coast."

I quickly pulled out my *Lonely Planet USA*, book since I couldn't look online.

"The Auto Club said it would cost nine hundred dollars to pay for a one-way rental car fee."

"I don't think so," I said. "You need to shop around."

"I can call around to the various car companies," he said.

When Jeff returned home from work, I said, "You need to call your dad. They need help planning their trip. By the way, what is the Auto Club?"

"Triple A, I think," Jeff said. "I remember them going there when I was a kid to get directions and maps for our vacations."

Jeff picked up the phone and called his dad.

"Hey Pop, how's the trip planning going?" he asked. "Share told me the rental car company wants to charge you nine hundred dollars. That doesn't sound right. You need to do some comparison-shopping online. Try Orbitz or Travelocity."

My father-in-law seemed timid about the idea and navigating the Internet.

"I can just call," he said.

"But online is much faster," Jeff encouraged him. "You'll get quotes from all the agencies at once. You just plug in where you want to go from point A to point B and the dates and it will list various options. And I heard you were told to fly into Astoria instead of Portland. Remember, the Auto Club person is not a travel agent, so I wouldn't necessarily take their advice."

"We just went down there to find out how long the drive time would be," my father-in-law said.

"You can do that online too," Jeff told him. "Just go to Yahoo maps and type in your departure address and destination and it will give you the time, distance, and directions."

Listening to the conversation, it sounded so ironic and hypocritical of us to be suggesting the use of technology but choosing to live without it. My father-in-law and mother-in-law are authentically unplugged. They prefer old-school methods -- newspapers, very little TV, minimum cell phone usage. But with our encouragement over the years, we got them hooked up -- new Apple laptops, iTunes accounts, wireless Internet.

Like most from their era, they are a bit intimidated by technology. There's a big learning curve and it can be frustrating and time consuming and may not seem worth the effort when their old methods work just fine. Things that we couldn't imagine living without, they couldn't be bothered with most of the time. We kept saying, "Once you retire and have more time, you can go to the Apple store and take classes on how to use your computer."

I'm beginning to think maybe that they had it right in the first place. Maybe their time could be better spent working in the garden or getting sucked into a good book. For me, one reason for keeping up with technology is to bond with Katelyn. I'm sure by the time she's in high school there will be some new fancy way of communicating. I want to be able to share that with her and relate to her. But I also want to teach her that there are choices -- alternative ways to do things, like using a real map instead of Yahoo Maps, because what if your iPhone dies while you're trying to find out how to get somewhere?

Day 131: My personality has definitely changed since unplugging. For one, I've become a hugger. I'm more appreciative of the friends that I have. Those who value my friendship will go the extra mile to reach me -- even when it's not convenient. I'm finding out who my real friends are. Those who can't be bothered by making the extra effort have dropped off. Sad, but true. I enjoy the one-on-one time with friends now, more than group outings. One thing that I've learned in life is that doing things that are not acceptable in mainstream can teach you who your true friends are.

The other change I have noticed is that I've become shy -- something very out of character for me. I've always been a social butterfly. Being unplugged, I have become self-conscious of my alternative lifestyle and people's disapproving reaction. I cringe at the thought of meeting someone new, knowing that inevitably, "I'll send you and an e-mail" will surface and I will have to divulge my technology-less lifestyle. I've learned to work around it and keep it to myself when I can, "Can I call you instead?" It's not that I'm not proud to have given it up. It's just that unless you've walked in the shoes of someone who has unplugged, you won't get it. People don't understand and think it's weird, rude, and inconvenient.

Author's note – *this is why I chose to turn our story into a book -- to help people understand the benefits and be understanding and supportive of those who choose to unplug.*

***

As I sit at my regular hang out, the Los Gatos Coffee Roasting Company, I can't help but be amused by how much my life has changed in the last few months. It's funny to me that I used to bring my laptop and take a seat at the bar and type my articles, surrounded by tables filled with laptop users, the noise of coffee grinder, and the smell of roasting coffee beans.

Just on the other side of the window, sitting outside, it's a different world. With book in hand, I indulge in some juicy literature surrounded by fresh air, sunshine, big and small dogs, and sweaty cyclists. There is a definite divide here that I had never noticed. I've crossed over to the outdoor/animal-lover crowd, which is probably what I've always been, but I was too clouded by my need for technology to see it.

Day 132: Katelyn and I are sitting on beach chairs in the garage with only a hanging bulb to illuminate the darkness.

"Let's open the treasure chest," I announce. I'm sure Katelyn envisions jewels and gold. Wiping off the thick layer of dust, I'm transported back to my childhood. The brass lock is rusted. With no idea where the key might be located, I use a screwdriver and hammer to break open the chest. I unlatch it and lift the lid to reveal mounds of handwritten notes.

"That's it?" Katelyn states disappointed.

I start sifting through my "treasures." This is more of a time capsule than a treasure chest. Katelyn is eager to dig through the pile -- convinced that she'll uncover the "goods" at the bottom. She reaches for some dried flowers.

"Careful," I warn her. "Those are very fragile. They were from my prom."

"Oh, did you wear a fancy dress?" she looks at me with bright eyes.

"Yes." She knows what a prom is thanks to Disney's *High School Musical Three*.

Memories flood back to me. My life in a box. I discover a lock of hair from when I was six -- same age as Katelyn is now, old photographs of me as a baby, lots of art projects from junior high, report cards from first grade through high school, a scathing note from my typing teacher who was always mad at my for typing notes to my friends instead of doing the assignment, my track shoes and fifteen blue ribbons I'd won in track meets at Eureka High School, schoolwork from first grade, a *beta* video of my eighteenth birthday (we were so hi-tech back then), and a junior high newsletter announcing that our school was getting computers. Most of the box is filled with letters I'd received from my Eureka friends after I'd moved to Fresno.

Katelyn's curiosity in my past amuses me. I enjoy sharing my history with her.

"This is fun Mama," Katelyn says rummaging through the chest. "I like seeing what you were like as a little girl, what things you liked."

Kissing her head, I hope her life can fill a box with wonderful memories. Two hours had passed, as I'm sucked back into the present moment and realize it's bedtime. I close the chest and leave with a sense of gratitude for a life well lived.

That night my dreams are infiltrated with friends from the past and I awake with a smile as if I'd had the opportunity to relive my years from first through twelfth grade, chatting with friends along the way. Where are they now?

***

My visit to the Conscious Living Center has inspired me to continue to explore my spirituality, but closer to home. I'd driven by the Self-Realization Fellowship building several times in Los Gatos and have always been intrigued by it, so I thought I'd see what it was all about.

The sign out front said the services were at eleven on Sundays, so I showed up a few minutes early. As I swung open the wooden door, I peered inside to see two women seated quietly. One woman put her finger to her lips to inform me of the silent time, but motioned me to come in. She whispered that it was meditation time. The doors to the main sanctuary were closed. She invited me to wait downstairs. I asked to peruse the literature in the lobby instead. Flipping through brochures, I learned about the founder, Paramahansa Yogananda, who encouraged his followers to live the life of Jesus' teachings combined with yoga and meditation -- East meets West religion.

Once the meditation time was over, the doors opened and I was invited to take a seat in the sanctuary. The lighting was dim. Rows of chairs faced the front of the room, which was adorned with paintings of six people, whom I assumed influenced the Self-Realization Fellowship. One I recognized. It was Jesus. I was ignorant as to whom the other five were, but later came to learn they were Lahiri Mahasaya, Mahavatar Babaji, Bhagavan Krishna, Swami Sri Yukteswar, and Paramahansa Yogananda.

The service consisted of cosmic chants, a sermon that focused on living a life filled with positive thoughts rather than negative, such as "Look for the good in everything." A bible scripture was read, songs were sung as the leader played an instrument that I was unfamiliar with. As I gazed around the room, inspecting the congregation, I didn't find a common thread. There were all ages and ethnicities in attendance.

At one point during the service we paused for ten minutes of meditation. I am familiar with mediation having done it at the end of yoga classes offered at various gyms throughout my life, so I mimicked others and sat erect with palms facing up, shoulders back and eyes closed. Naturally, my thoughts drifted to things I needed to get done during the day -- buy some chocolate chips to make cookies for Katelyn's teacher, go to Target to pick up my film (hope I loaded it correctly), figure out what to make for dinner. Then I remembered that meditating was supposed to clear your mind so I tried to focus on my breathing -- *in and out, in and out*. After a few minutes of silence, I peeked my eyes open and looked around the room.

It was surreal to be sitting in a room with over a hundred people and have it be silent. Something I had never experienced. Quite different from the meditating I've done lying on the smelly gym floor with a handful of moms. I could feel a sense of calm energy. As I closed my eyes again, for what seemed like an eternity, I felt as if the others might somehow be able to hear my thoughts, so I tried to clear my mind by using a method a yoga teacher had once taught me.

"Let your thoughts come. Recognize and accept them, but then see them float past like a cloud rather than focusing on them," she had told the class. A thought that entered my mind was lunch. As I tried to let a vision of a bean burrito float away, my stomach growled...loudly...breaking the silence for those seated next to me. *How embarrassing!* The leader gently guided us back to the present moment to continue our worship.

The service ended with several blessings, hands raised, and the congregation loudly chanting in unison, “Om.” A final “Om, shanty, amen” and we filed out. I left there feeling as if I had witnessed something meaningful. I wasn’t convinced that it was the right religion for me, but I was thrilled with having been given the insight as to how others live. It gave me respect for an alternative way of living. When I was younger I used to think that adventure was defined by doing something that was frightening or dangerous, like bungee jumping, mountain biking, snowboarding, mountain climbing. Now, I’m thinking that adventure is trying something new that is not what you would typically do. I now like the idea of becoming more well rounded by seeing life through the eyes of others. It’s interesting and educational and entertaining. It keeps you from being narrow-minded. It makes you less judgmental. I appreciate gaining a new perspective on things by having experienced it.

I hope that I no longer say, “I have no interest in X, Y, Z,” but rather, “I don’t know if I’d like it, but I’ll give it a try.”

***

When stressed, the norm for me is to stand in front of the cupboard or refrigerator -- door open wide -- scanning for some comfort food and shoving it down as quickly as possible, but this time something different happened. I grabbed the garden hose instead and spent a quiet moment feeding my plants and enjoying the calmness of the garden. Then, I pulled my lounge chair onto the grass, leaned back, let the sun warm my face and opened my "book of the week." The thing was, this happened intuitively. I didn't even think about it until afterwards. Technology promotes the need for the quick fix. So, it made sense that when I was plugged in, I would also grab food as a quick fix for my emotional needs. Since unplugging, I've learned to slow down and feed my soul through more nurturing activities, leaving me feeling calmer and more fulfilled. This is something I've been striving for my entire life, but was only able to realize by giving up my technology -- a true "ah-ha" moment.

***

After partaking in the last summer concert in the park in Los Gatos, we returned home with bellies filled with wine, cheese, strawberries and chocolate and souls filled with love, music, and friendship. Jeff tucked Katelyn in while I snuggled up on the couch reading a book.

"She fell asleep in the middle of Cinderella," Jeff said as he sat beside me on the couch.

I was feeling particularly drawn to him at the moment. He had sacrificed an afternoon of work deadlines to splash in the pool with Katelyn and partake in the concert festivities. This meant the world to me -- a selfless act. Jeff was also leaving on a business trip the following morning -- off to Texas for three days. I pulled him close to me and kissed him. I wrapped my arms around him and pulled him on top of me, feeling his body responding to my advances as he pressed against me. It is possible that we had made love on the couch before, but I couldn't remember when...long ago. Normally, our love was confined to the bedroom, behind locked doors and away from the potential prying eyes of a six-year-old.

"Are you sure Katelyn's asleep?" I whispered softly into Jeff's ear.

"Yes, she's out," Jeff replied.

With gardening on my mind, I said, "We should make love outside in the yard sometime."

Without hesitation, Jeff responded, "Why not right now?"

I thought about it for a split second and replied, "OK, grab a blanket."

We tiptoed into the yard, protected from prying neighbors by a six-foot fence. The warm and dark summer night was inviting. Jeff spread the blanket on the grass. We giggled quietly as he inched my dress up my thighs. I lay on my back listening to the crickets and gazing up at

Jeff's face with a backdrop of twinkling stars behind him and the half moon illuminating his smile. Feeling the soft breeze against my skin was exhilarating. We were creating a bond with each other -- as well as nature. I was just praying mosquitoes weren't obliviously attacking us during our backyard tryst. The night was magical -- filled with a sense of daring adventure mixed with romance and intimacy. At that moment, I did not miss the TV one bit.

Day 135: Katelyn's in the back seat of the car -- her long blonde locks blowing in the wind.

"Mama, I want to cut my hair and donate it to the Locks of Love," she blurts out.

"That's a generous idea, but are you sure you want your hair short?" I inquire.

"Yeah, I don't want to be like *Hannah Montana* anymore. I've changed a lot. I need a new look. I want to look older. Plus, my long hair gets caught in my armpits when I'm playing on the playground," she said twisting her locks around her finger.

Could it be that Disney's grip on her was slowly slipping away?

"But if you cut it, you won't be able to put your hair in a bun for ballet, or braids or pigtails," I say.

"That's OK. I want a bob. I just want to look different," she said.

As a gesture of support, I offered, "Want me to get my hair cut short too? I don't mind."

"Mommy, you don't have to copy me. You can keep yours long," she said.

Day 136: Back to school night. Parents are crammed into the child-sized desks and chairs as they gaze proudly at their child's artwork, writing, and drawings that are posted around room. After the teacher fills us in on the students' daily schedule and yearlong curriculum, she welcomes us to the whiteboard to sign up for volunteer opportunities. The options include being a room mom, clay assistant, crossing guard, or art helper, or doing tasks such as photocopying, filing, book ordering, or working in the class. I'd love to be room mom, a liaison between teacher and parents, but a thought dawned on me…I would need e-mail. *Damn it.* This is really cramping my style.

As we were walking out, posted outside of the office were dozens of additional volunteer opportunities for the school: garden coordinator, fundraising, walk-a-thon chair, talent show coordinator, yearbook editor. I opted to work in the classroom and garden, be the clay mom and help with stapling. It would have been nice to be even more involved and get to know some of the other moms better, but I would take what I could get at this point.

In the multipurpose room, the principal gave a speech about the quality education that our children would be receiving, introduced the teachers, talked about fundraising, discussed the Home and School Club (our version of a PTA), and ended by saying, “I’ll be sending a newsletter out every couple of weeks with info about the school and, as you may have discovered, we are a ‘green’ school so we will be sending the newsletter via e-mail from now on to save some trees. So make sure we have your current e-mail address.”

A knot in my stomach.

***

My mother-in-law called to inquire about purchasing a digital camera.

“I want to get one I can use with my Mac,” she said. “I think you can make a slideshow or something. Do I buy that at the Apple store?”

“I don’t know if Apple sells cameras at the store since the iPhone now has a camera in it,” I told her. “Try Target. They have a large selection. Wait a minute, you’ve never owned a digital camera?”

“No, we just use throw-aways,” she said.

“But I’ve seen you use a real camera. I didn’t realize it was a film camera.”

“Yep. Oh, I don’t know where there is a Target, but I’ll figure it out.”

“How can you survive without shopping at Target? I asked, in shock. “Where do you shop?”

"Albertsons," she replied. "We don't need much"

"I'm at Target every other day it seems," I admitted. "Maybe it's different with having kids."

"You must think we are so primitive," she replied.

"No, you're inspirational," I said.

Day 142: When Katelyn was about two-years-old, I thought I hit the jackpot when I found a stack of Disney princess movies at a garage sale for fifty cents each. Even though I remembered watching them fondly as a young child, as an adult I was horrified to realize that every movie had such dark undertones.

- Cinderella had to endure her wicked stepmother
- The evil queen wanted to rip Snow White's heart out
- Ariel gets her voice stolen by a vicious octopus
- Pocahontas' entire tribe is in danger of being wiped out by the white man
- Mulan fights in a war
- Sleeping Beauty is cursed to die
- Belle gets locked in a dungeon by a terrifying beast
- Jaffar makes Jasmin his slave

And why do they all have to get married to be happy? I ended up giving the movies away.

***

While breaking in my old typewriter, I glance at the bookshelf and got a glimpse of my "past life" with books dedicated to my technological lifestyle, *HTML for Dummies, CSS, Photoshop, QuickBooks, XHTML, How to Start an Online Business.*

Now I'm reading books about gardening, cooking and sewing. Am I turning into a 1950s housewife? Jeff's going to walk in the door after work one day and be greeted by me wearing a cooking apron that I sewed myself and holding a casserole made with veggies that I grew in our backyard garden. Maybe we need a chicken coop to collect fresh eggs.

***

"I made a new friend on the playground today," Katelyn said as I picked her up from school. "Not to be mean, but she's fat. But I played with her anyway because she was nice."

I thought I was going to have to call 911 to reattach my jaw that was on the floor. We live in a commercial-free zone. I've always told Katelyn "people come in all shapes and sizes." Even without the influence of TV, Internet, radio, some girls still feel the pressure to be thin and think that being fat is not socially acceptable…and at such young ages. Probably time to ditch my *People* magazine subscription -- not sending the best message.

That night when Katelyn was helping set the table she asked, “Are you going to have wine or water with dinner?” So that she could set out the appropriate glass. I cringed. Again, setting a bad example. Probably best to keep my indulgences in wine and *People* until after she’s in bed -- or while she’s at school? A mid-morning mimosa? Yeah, that would lead to a *very* productive day.

***

I had an epiphany. I might not be who I think I am. I thought I could not live without technology, but I can. Technology has been such a huge part of who I am -- defining my career, social life, relationships, education, interests. I’m realizing that I can be someone other than who I thought.

If this is true, does that mean I can take any of my negative attributes and change them by telling myself they aren’t true. For example, I have battled a terrible sweet tooth for as long as I can remember. So, what if I just start telling myself that I don’t care for those things? Eventually, will my mind and body start to believe it? Can I redefine who I am? Maybe I really am *not* afraid to jump off the high diving board, maybe I really *do* like roller coasters and sushi. Who knows? It’s so easy to get into habits and ruts that define us and, after so many years, it's just too hard to break free, so we accept it as "that's just me." Maybe it's not.

***

In my quest to find alternatives to the treadmill, I headed to the gym for a Pilates class. When I got there, I discovered that I had misread the calendar and instead it was a "strength and stamina" class. The instructor convinced me it was worth a try, plus I had driven all the way down there. It seemed a waste to turn around and go home. I knew I was in trouble when she cranked up the volume on the *Black Sabbath* CD and was yelling commands through her head microphone. *Jab-Jab, upper cut, sashay to the right, plié, karate kick.*

The music was thumping so loud it gave me a headache. The steps were so complicated. I think you would have needed a dance degree to follow them and I am so uncoordinated. This was torture -- not how I had intended to spend my morning. I had envisioned some gentle, yet challenging poses, not boot camp with Ramboette.

Of course I was situated the furthest from the exit. To leave the class I would have to scoot past 20 sweaty people with my mat, weights and stretchy band to return to their proper places. Too embarrassed to make a scene, plus there were two grey hairs keeping up. I didn't want to look like a complete wimp.

As I painstakingly stuck it out, I began to realize -- no matter how difficult (or torturous) something seems (like living without technology), if you hang in there, you eventually get past the discomfort and might actually realize you enjoy it…or not, but either way you'll feel good having given it a good shot. After all, the things that we find most challenging, usually end up being the most rewarding and help us grow.

Day 145: With the warmth of the sun, consistent hydration and a little tender loving care, our garden has flourished. The lettuce has grown from fledgling seedlings to hearty heads, twelve to be exact. The basil, mint and rosemary are reaching up as if to shake hands with Mr. Sun to thank him for his nurturing rays.

With a new cookbook, *Green Food*, I flip through the pages and land upon a recipe that calls for lettuce and basil, a perfect excuse to harvest our crop. I bend down toward the earth and pluck leaves from the lettuce. Katelyn watches intently, squatting next to me, her eyes squint in concentration. Jeff records the moment for prosperity -- fumbling with the 35mm camera. Katelyn and I hold up our bounty and flash a proud smile. In the kitchen, I whip up my masterpiece and serve it to my famished family.

For the first time in her life, Katelyn announced, “Mmm…I like lettuce.” No TV for tuning into the Food Network. No Internet to search for recipes. Just a book and some dirt to convince my daughter that salad can be a good thing.

Month Six: Can You Hear Me Now? No

"My cell phone is my best friend. It's my lifeline to the outside world."
--Carrie Underwood, Country Singer

Day 150: Ever since Jeff's niece was in elementary school and did an overnight camping trip to Catalina with her dad as part of YMCA's Adventure Guides, I knew that if we had a daughter, Jeff would have to join.

Since the indoor rock climbing never came to fruition as a weekly daddy/daughter activity, the YMCA Adventure Guides seemed like an alternative that might be more feasible with its monthly meetings -- special time for them to bond over bonfires and s'mores. Not only would Katelyn make new friends, but also Jeff, who was lacking a social outlet since unplugging.

At school one day, I asked my friend Sage, "Your daughter Sandy does Adventure Guides, right? I'd like Katelyn and Jeff to check it out. Do you know how we find out when the next meeting is?"

"I know they plan everything through their Yahoo group, *buuuut* since you're not online I'll ask my husband Steve about it," she said shaking her head.

She called that night and left Steve's cell phone number so Jeff could call him. The next day at school, she showed up with a print out of the time and address of the next meeting in two days. That evening Jeff was reluctant to get involved since he was so swamped at work.

"What if there's an outing and I'm out of town on a business trip? I don't want to disappoint Katelyn," he said with concern.

"It's once a month. You can fit that in," I said.

I had to dial Steve's number and hand the phone to Jeff, knowing he would never get around to it. After chatting, he hung up and said he'd go.

"You have three choices to find out where the meeting is, since we only have the address and no Yahoo maps. One, call Steve and ask for directions or carpool. Two, call the guy hosting the meeting and ask for directions. Three, ask our neighbor Ned for directions or carpool with him, since he and his daughter are in the same group," I told Jeff.

I went to fold laundry in the bedroom and when I returned to the kitchen, there was Jeff, a city map sprawled across the kitchen table. I began to laugh hysterically.

"You are so funny! What is it with guys not wanting to ask for directions? Why don't you just call?" I said.

"I don't need to. I found it," Jeff said proudly.

The night of the meeting, a knock on our door. Ned, our neighbor, and his daughter, came by to see if Jeff and Katelyn wanted to carpool to the meeting. Ironic.

Katelyn came bounding in the door after the meeting at eight-thirty.

"Mama, they had a full-on Indian drum and money pouch to put our money in for the end-of-the-year pizza party. And I had to tell everyone my Indian name, Tusayana."

She babbled non-stop about the girls she met, the art project they did and the ceremony.

"So, what did you think?" I asked Jeff

"It was fine, but the other dads already know each other, so I kind of felt like an outsider," he said.

"You'll get to know each other soon enough. Seems like Katelyn really enjoyed it."

"The other problem is that they do everything online -- send meeting invitations, pay for events, send the calendar. How will I get around that issue?"

"Welcome to my world. You'll figure it out," I reassured him, not entirely convinced that he would.

"Did you want me to do this just so I would have to learn how to deal with being unplugged?"

"Of course not! But you're about to find out just how challenging it is."

Day 152: As Katelyn, Lesley, and Sandy whirled around the backyard at Sage's house, Maddy, Sage and I sat comfortably inside, engrossed in the Food Network. Sometime during our conversation, inevitably my life unplugged became the topic of discussion. I'm not sure how it came up, Maddy recommending looking up a book online, Sage rousting me for not being on the school Yahoo group, but the next thing I knew they were on me like bees on honey.

"You are such a creative, multi-faceted person, you're really limiting yourself by not being online," said Maddy.

"I just don't get it," Sage added. "Fine, do without TV, but not the Internet. You're just making things hard on yourself."

I found myself smack in the middle of an intervention. My friends continued to convince me that I needed to plug back in and I felt the need to defend myself and informed them of the benefits of living without technology.

"Katelyn has become more independent. I have become more relaxed. We are spending more time together as a family. Jeff and I are reconnecting. I'm reading, gardening, sewing, rather than being tied to my computer or TV. I'm setting a better example for Katelyn and being more available to her."

They had enough of the nonsense. It drove them crazy that I was living such an alternative lifestyle. They wanted their friend to return to *normal*.

Day 154: The date was quickly approaching. I hemmed and hawed over the decision. *Should I or shouldn't I?* The event that had me so torn…the *Hannah Montana* concert that was about to take place 15 minutes from our house at HP Pavilion. If I took Katelyn she'd think I was the coolest parent in town, plus it would be a good bonding moment. But was it appropriate? She's only six. Did she need to go to a rock concert at such a young age?

Jeff was concerned. *What if the concertgoers are scantily clad?* It wouldn't be the best impression for Katelyn to be exposed to. My final decision -- you only live once. When would *Hannah Montana* be in concert so close to our house again? I had to take her. She would treat me like a rock star for treating her to such an event. That night I told Katelyn, "Guess what? I have a surprise. I bought *Hannah Montana* tickets."

She let out such a loud, high-pitched scream, I was afraid the windows would crack. Unable to contain her excitement, she thrust herself around me and hugged me tighter than a vice grip.

The night of the concert, I let Katelyn be my stylist. She chose jeans, a sequined T-shirt and high heels for me. She pulled my hair back in a clip to showcase my over-sized silver hoop earrings. The final touch, a dab of lip-gloss "carefully" applied (it's actually supposed to go *on* your lips -- not also around them). For herself, she chose a sparkly pink T-shirt with matching sequined scarf and leggings.

Upon arrival, we were ushered to our seats…behind the stage. Not back stage, but behind. We'd be looking at *Hannah Montana*'s back. Considering the ridiculous amount of money I dropped on the tickets, this wouldn't do.

Without Internet access, I had called Ticketmaster (I found the number on a flyer that was mailed to us promoting the circus -- for once, thankful for junk mail). I was unable to look up the seating chart online, so I put my faith in the automated system.

We marched to the ticket booth and demanded better seats. "There are no better seats," the ticket cashier informed us. I requested to talk the manager and held Katelyn on my hip as I explained our desperate situation through the speaker as the safety glass separated us. Maybe the glass was to prevent irate concertgoers from strangling the ticket sellers for giving them crappy seats that they spent half of their retirement on. Surely, once he saw Katelyn's adorable face, he would have to take mercy on us. He did.

He slid the new tickets under the window. Not front row but a side view from the balcony. An improvement…so we thought. Miley's brother opened for her. How convenient, but not a good fit for the audience. As Mr. Cyrus rocked out -- screaming grungy lyrics at the mostly under eight years crowd, we were blinded by stage lights that shone directly in our eyes. Katelyn covered her ears with her hands, trying to muffle the too loud, heart-thumping songs that blared from the over-sized speakers. I held my hands up to shield our eyes from the spotlights. After sticking it out for a few moments, I tried to squish some earplugs into Katelyn's ear, but had been blinded by the light and couldn't see what I was doing. Katelyn squirmed uncomfortably.

"Let's go," I yelled. She shook her head in agreement. We waited in the lobby, a reprieve from the over stimulating concert. An HP Pavillion employee with a microphone headset passed by. I flagged him down and begged for better seats yet again. After a half-hour of being concert refugees, we sat on steps waiting for some reprieve. No one showed up. We slunk back to our balcony seats when the opening band left the stage. As I gazed around the concert, I saw young girls waiting anxiously for the main event, dressed in authentic *Hannah Montana* gear -- mini skirts, converse, sequined tops with vests. Our tickets said Miley Cyrus Tour and the signage hung strategically throughout also reiterated Miley Cyrus. The souvenir booths sold Miley Cyrus T-shirts. *Hannah Montana* was nowhere to be found. I prepped Katelyn, "I'm not sure if she'll be wearing her *Hannah Montana* wig." Surely she'd sing her hit song, "The Climb."

The house lights dimmed, the stage lights flickered and from the middle of the stage arose Miley Cyrus through a trap door. She donned a mysterious "boxing" robe with hood covering her face down -- singing a Miley Cyrus song we were unfamiliar with. The crowd burst into hysteria, bouncing on their seats at the sight of her. She flung off her robe and revealed mini black leather shorts and low-cut tank top that was even more revealing when she bent over to rock her lyrics and engage in head-banging antics. Katelyn's face beamed. My face froze in shock. *Uh, so not appropriate.* We stuck it out and were subjected to several more skanky outfits and unfamiliar rock (not cutesy pop) tunes. The wardrobe changes consisted of an ultra-mini tutu with "underwear" clearly showing, matched with a peek-a-boo bustier, a man's white button-up shirt sans pants (reminiscent of *Risky Business)*, but included a tux jacket with a long train behind, and at one point I thought she may have had a wardrobe malfunction when she strutted around the stage with skin tight black shorts that looked like underwear. The theatrics were entertaining as she flew over the audience on a wire and sailed through the air atop a vintage motorcycle and drove an off-road vehicle onto stage, but I couldn't help but wonder how her parents could approve of her wardrobe choices. I guess sex sells, even at age 16.

At one point she said, "Who's ready to party?" *Do elementary kids "party"?* After about 45 minutes, it was nearly nine on a school night, Katelyn said, "I'm surprised she can stay up this late. She's only 16." Miley yelled to the crowd of fans, up past their bedtimes, "Get up and dance. You can sleep at someone else's concert -- not mine!"

No "The Best of Both Worlds." No "Nobody's Perfect." No glam pop star. Hannah has a dark side. Was Miley feeling the need to rebel against her Disney icon and show her edgy side? I was utterly repulsed as she suggestively rubbed the chest of her (much older) guitar player while crooning. *Eww!*

A couple of dance moves included running her hand from her neck, across her chest and stomach and stopping at her crotch with a hip thrust. Can you say *exploited*? Katelyn had gone from elated to disappointed. Not what we had anticipated. I was horrified and too embarrassed to reveal my grave mistake to my friends the next day and would never look at *Hannah Montana* the same again.

Day 155: My mother-in-law called. "Share, you're going to be so proud of me…I took my digital camera to CVS and had them put my photos on a CD," she beamed through the phone.

"That's great, but why did you choose that route?" I asked.

"That's what Kiely (Jeff's eight-year-old niece) suggested."

"That's really good to have as a backup, but, just so you know, you don't need to do that. You can just upload them straight to your computer. And if you want prints, you can order them online and have them mailed to your house."

"Really?" she laughs. "That's something. And what is iPhoto and how do I get it?"

"It comes with your computer. When you plug your camera in, it will open automatically and allow you to view your images," I informed her.

Hours later she calls back.

"I created my own albums in iPhoto. What a kick," she giggles. "But, I'm not sure how to get the photos off the camera and onto my computer." *If only I lived nearby to help her.*

Day 159: Jeff and I sat side-by-side in the home office. I was pounding the keys on my typewriter, jealous of Jeff's ease of working on his laptop, as we shared a bottle of Pinot Noir while Katelyn snoozed away in her princess bed after a hard day of swinging on the monkey bars and making a masterpiece with finger paints. In need of a break, I straddled Jeff and kissed him passionately in his desk chair. Then hopped off and went back to work.

"Wait a minute," Jeff said, pulling my rolling desk chair towards him. "You can't just kiss me like that and then stop."

He pulled off his T-shirt and began to undress me.

"You know, we haven't broken in the Prius," he smirked.

I smiled and grabbed the car keys and threw on my robe.

We shuffled barefoot to the car. Moving Katelyn's car seat to the front seat, we tuned in the radio for some mood music. Jeff sat in the backseat with me on top.

"I think we should lay down so the neighbors don't see," he warned. "And turn off the radio because the light from it is too bright."

Jeff lay down. As I tried to position myself, my left knee dug into the seat belt. We giggled and groped.

"If the car's a rockin', don't come a knockin,'" I joked.

Neither of us could relax enough to get the job done, so we whisked back into the house to the comfort and security of our bedroom.

The next morning, I hopped in the car and laughed when I saw Katelyn's car seat in the front seat.

Day 161: Jeff's coworkers are back from their the honeymoon and invited us over for dinner.

"Tell them we'll bring a salad from our garden," I yelled to Jeff as he accepted the invitation.

After indulging in the nearly homemade pizza, a la Trader Joe's, pasta salad, garden salad and ice cream sundaes for dessert, we were asked to partake in some after-dinner entertainment...Guitar Hero.

"This was the whole reason we bought a TV," said George.

We were Guitar Hero virgins.

I told Katelyn, "This is that guitar video game we saw the kids playing at the community center."

Once the peripherals were plugged in, we assumed our positions. I was the singer, Jeff was on drums, Brenda and George on guitar, Katelyn was my backup singer.

Throughout the evening we rotated instruments as we belted out Beatles tunes. Now this was a good use of technology -- bringing friends together through fun and laughter. I thought the Wii doubles tennis was fun, but Guitar Hero was a group effort with a common goal, not a competition. Turned out to be a fun night filled with good food, good friends and lots of laughs.

Day 162: "Mama, today at school, Justin, you know the boy who sits next to me in class, lifted his shirt and showed me that he had a Nintendo DS stuffed in his pants," Katelyn snitched.

Luckily he wasn't showing her the other thing stuffed in his pants!

"Why didn't he have it in his backpack?" I questioned.

"I don't know," she said. "But if the teacher saw it, she would take it away. He plays with it at recess."

"He doesn't play on the playground with the other kids?"

"No, he just sits by himself and plays his DS."

That's so sad. Does his mother know about his secret?

***

"Quade called," I said to Jeff. "Apparently, Vinne is in town from Boston for the night. I guess he sent out an e-mail, but obviously you didn't get it. Quade called to see if you're going to meet up with them."

"I'll give Vinne a call to see what's up," Jeff said, reaching for his iPhone.

"Hold on. You can't call him on your cell," I scolded him.

"I know. I'm just looking up his number," said Jeff.

"Wait a minute. You still use your iPhone as your address book?""

"Yeah, so?"

"*Hello*. We are unplugged," I said irritated. "You can't use an electronic address book. How did you get away with that for the past five months? Why do you think I have to lug around my huge Day Runner. You're such a cheater! Five months!"

"Well, I don't have an address book."

"I'll buy you one tomorrow. I can't believe you. What else are you doing? Surfing the web at work?"

"No!" Jeff said. But I wasn't sure I entirely believed him.

Day 164: As Katelyn and I careened down Highway 101, following a monster shopping spree at Ikea, I glanced down at my blinking dashboard. A red exclamation point lit up and flashed relentlessly as if shouting at me. *What is that?* I panicked. My low gaslight also blinked. *Oh shit! I forgot to get gas before making the 30-mile trek to Palo Alto on an empty tank.* The thing about owning a hybrid is that you only have to get gas every couple of weeks so it's not something I pay much attention to. *Blink, blink, blink* -- the red light flashed. A four-lane highway was not a good place to have car trouble.

"Katelyn, I need to pull over. The car is acting funky," I said, trying to keep my voice as calm and smooth as an early morning lake. I pulled off at the next exit, which was unfamiliar to me. As we rolled off the highway, we entered the intersection onto the expressway and the Prius decided to take its last breath. It died in the middle of the intersection! I hurriedly turned on my hazards and began apologizing (to myself) for blocking traffic. “I’m sorry, I’m sorry,” I muttered as cars veered around me honking as they passed. Opening my car door to exhibit engine trouble, I suddenly realized we were in serious danger as the oncoming traffic had a green light and they sped towards us. *Do I ditch the car, grab Katelyn and run?* I tried to shift to neutral but the gears were jammed.

As I sat in fear, a man appeared at the driver window with another at the passenger window. Before I barely had time to explain, they were pushing my car to the side of the road. I felt like a princess being rescued by two brave knights. I quickly thanked them -- so they could get out of harm’s way and on with their lives. “You definitely win good Samaritan of the day award,” I told them, wishing I had their addresses to send them a thank you.

On the side of the expressway, I pulled Katelyn from the car and positioned her against the concrete guard wall. Retrieving my cell phone from my purse, I called our insurance company -- Farmers. While on hold, Katelyn said some wise words of advice,

“It’s a good thing you have your cell phone, Mama. Looks like you do need it after all. Maybe living without our computers, TV, and stuff isn’t such a good idea.”

As if things weren’t bad enough, now I’m getting a lecture from my six-year-old.

"We'll be fine," I reassured her. "I'm calling Farmers."

Katelyn replied innocently, "Why would a farmer come help us with our car?"

A chuckle was just what I need in this time of stress.

While on the phone, a tow truck pulled up from AAA. I explained that I didn't have AAA. Seeing Katelyn's uneasiness, he took mercy on us and agreed to help us free of charge since Farmers would take an hour to arrive.

"Do you have some cash? I can go get some gas for you," he offered. The thought of being robbed roadside never crossed my mind. Luckily he wasn't a maniac preying on broken down car victims.

"Actually, I don't think I have any cash (and I couldn't exactly give him my bank card). I scrounged up two dollars. He took it, and returned from his truck with a gas can.

"I think I have a gallon to get you to the next gas station. I'll follow you to make sure you get there OK," he said, pouring the gas into my tank. I swear I heard the Prius breath a parched sigh of relief. A push of the start the button and the Prius came back to life. But the danger lights were still illuminated. We carefully drove to the nearest gas station, five blocks away.

Suddenly, Katelyn whines from the back seat, "I gotta go potty!"

"We're almost there. You can go at the gas station."

"But I'm going to poop my pants!" she yelled.

"No, no -- hold on -- we'll be there in a second."

"Oh no, I have to pee, too! Mama, I'm going to pee in the car! I can't hold it," she jiggled in her seat.

"Just hold on!" I yell in a panic.

We pulled up to the gas station with the tow truck behind us. I quickly unbuckled Katelyn, hurriedly and profusely, thanked the man who helped us and vowed to become a AAA member, then whisked Katelyn to the cashier.

"Where's your bathroom?" I asked as Katelyn was bent over trying to hold her poopie and pee-pee in.

"We don't have one."

Just our luck.

"Where's the nearest one?" I panicked.

"Not sure."

I scanned the parking lot and surrounding buildings -- Walgreens. We raced across the blacktop and flung open the Walgreens public restroom door. Perched on the potty, Katelyn unleashed the worst bout of diarrhea her little body has ever seen. *Thank God she didn't poop in our new car!* As we returned to our car parked at the pump, the AAA guy was still there.

"I just wanted to make sure you were OK," he said, obviously having seen us race across the parking lot.

I filled the tank, gave the car a minute to let its meal settle, and the lights went dark. Flashing two thumbs up at our saint, I waved him off.

Day 165: Dropping off my film at Keeble and Shuchat in Palo Alto, I walked past a homeless man holding a sign, "Help us grow a Homeless Garden." I couldn't resist asking his story.

Turns out, he was fired eight years ago as an engineer at Lockheed Martin. Upon losing his job, he also lost his house and his wife. Since then, he's been living in his car and panhandling.

"I want to start a garden to feed the homeless," he told me, his dirty, wrinkled face showing signs of a hard life. "I started one eight years ago, but it became corrupt. So, this time *I'm* going to run it. I'm running for City Council to make sure it happens."

"Don't you need an address to run for City Council?" I asked.

"I have one…telephone pole 1139," he replied, pointing to a pole nearby. "That's what's on my voter registration card. The milk crate has become the best political pedestal."

He requested I go to smartvoters.org to learn more about his campaign. Although I didn't offer him my spare change, I would have been happy to help his cause had I had access to the Internet. Even the homeless in Silicon Valley are online.

***

While pulling weeds at Hidden Villa, my manager David announced that we had to make the garden extra pretty because they had their annual fundraiser that weekend and the guest speaker was author Richard Louv. David said volunteers were welcome to attend the event.

If I had not gotten rid of the TV, I wouldn't have taken the time to read *Last child in the Woods* by Richard Louv. Had I not given up Internet access at home, I would not have been motivated to volunteer at Hidden Villa. Had I not volunteered at Hidden Villa, I would not have had the chance to *meet* Richard Louv.

I'm utterly amazed at how things come full circle once you start down the path that's meant to be.

Day 167: Fall has arrived. The swimsuits have been stored. The pumpkins purchased. Time to plant the veggie garden. Since it is a "Friendship Garden," we decided to invite some friends for pizza and planting -- a group of friends we've known since our kids were two months old. While Jeff slaved away on our gourmet dinner – homemade pizza, I invited the kids to do some planting. Our new addition to the garden would include rainbow chard, cabbage, white cauliflower, purple cauliflower, beets, onions, purple broccoli, and thyme.

I got a couple takers, but the kids seemed mostly interested I running around the yard instead. *So much for putting them to work.* Three kids each planted one plant, then were done. The main responsibility fell on Tina and me. She joked, "So, we have to work for our meal."

With plants in the ground and pizza out of the oven, we gathered around the picnic table for our meal that including a fresh garden salad, minus the mini snails I pulled off while washing the lettuce.

As is typical in Silicon Valley, the dinner conversation quickly turned to technology…Facebook to be more precise.

"I found a few school friends on Facebook and it was fun at first, but then we realized that now we don't have much in common anymore, so I don't really keep in touch with them," said Ella.

Beth added, "Apparently high school reunion attendance is down because now everyone is reconnecting on Facebook."

In her book, *The Facebook Era,* author Clara Shih discusses the reality of teens today. They could literally keep track of everyone they meet in their life through Facebook. She brings up the question, "Is there a limit to the capacity humans have for how many people they know and will it stop them from making new friends?" I posed the question to our dinner guests.

"Why would I want to keep in touch with everyone?" Tina said. We all agreed that it's the quality, not the quantity that matters. I inform my friends that I have apparently committed Facebook "suicide" by not being on it anymore.

As the night cooled, the adults moved inside to the living room for a game of Would You Rather. That night our kids had a pivotal change in their friendship. Three girls, three boys -- ages four to six. They've known each other since they were infants and have had numerous play dates, dinners, vacations together, but haven't ever quite clicked. They didn't form the close bonds that the moms have, but throughout the evening a new connection transformed.

Upon arrival, each child was playing alone or with only one other child -- some were feeling left out and bored. Tina tried to mediate, like usual, making sure everyone felt included, but the kids could never agree. The girls wanted to play jump rope, the boys racing games. We gave up and let them fend for themselves. By the end of the evening, I was beginning to wonder if my garden was *magical.*

The kids had chosen (on their own) to play hide and seek -- boys against girls. All of the parents were in awe as our children ran through the house and yard giggling and screaming. When the sky grew dark, they begged for flashlights so they could continue their game outdoors. Had we been plugged in, inevitably the solution would have been to put on a movie the kids could watch together...with no interaction. Without that as an option, the kids had more fun actually playing together.

***

We got the call. Our babies were ready to come home...baby kittens, that is. Katelyn and I rushed to the Humane Society to greet our temporary pets. Peering into the cage, we were greeted by one white fluff ball and one black one. The staff carefully put them into carriers, gave us some kitty food and we were off.

"What should we name them?" I asked Katelyn as they softly mewed from the carrier.

"I don't know yet," Katelyn said peering, into the carrier from her car seat next to them.

"How about Salt and Pepper?" I suggested.

"No, the white one will be Snowball. I'm not sure about the black one," she said.

"How about Spooky?"

"No, maybe Roses."

"Roses isn't really a name, but Rose is," I said.

"Are they both girls?" Katelyn asked.

"Oh, I'm not sure. We'll check when we get home."

"Oh, Mama, they are so cute. I think they like me," she swooned over them.

The kitties were six weeks old, which meant we'd foster them for only two weeks. Perfect. To accommodate them, we sectioned off half of the office since we were told to keep them contained. A bookshelf would suffice as a barricade in the middle of the room. The hardwood floors were covered with a sheet and the new kitty home included a bed, litter box, food and water.

"Welcome home kitties!" I said placing them in the new environment. They stayed hidden in the comfort of their carrier. I gently pulled them out and held them on my lap. Katelyn sat close by as I stroked their soft fur. They purred happily. As a child, I had grown up with many cats -- Oddball, Spooky, Scamp. Cats, to me, were always warm and comforting companions that loved attention, but didn't need it -- an independent, but dependable pet. Katelyn kept a cautious distance.

After deciding the sexes, Katelyn decided that the white one, which was a girl, would indeed be named Snowball. The black one turned out to be a boy.

"How about Jasper?" I said.

"Oh yeah, like the cat in the book we read at school!" she replied enthusiastically.

It must have come to me subconsciously through Katelyn's homework. So, Snowball and Jasper were welcomed into our home and hearts…for two weeks.

Day 169: Katelyn is sick. I am sick. Jasper our kitty is sick. And Jeff is attempting to take care of all of us while being frustrated by not being able to simultaneously get his work done, since he doesn't have Internet at the house.

For having a 102 fever, Katelyn is bouncing around the house full of energy. I, on he other hand, don't even have the energy to open one eye. I have created a cocoon for myself and intend to stay put. After spending the morning taking Katelyn to the pediatrician and Jasper to the vet, I desperately called Jeff to come home from work and then promptly collapsed.

The smell of musty cat food and fresh poop have infiltrated my office. Jasper scratches and Snowball bites. Only two days into it and I'm over the cuteness factor. As Jasper gasps for breath, I hope his meds kick in soon. We don't need this to end in tragedy.

***

My answering machine light flickers. I press the play button to hear, "This is Amelia from the Carlson Clay Club calling to let you know we're doing a training this morning at nine. If you could call me to give me your e-mail address, that is how we communicate."

Oh, yes. I signed up to be a "clay mom" at Katelyn's school.

Home sick, I wouldn't have made the training anyway, but I was a bit put off by the late notice. Returning the call that evening, I said, "I got your message. I'm sick so I couldn't make it today."

"What's your e-mail so I can include you on the e-mails?" she said abruptly.

"I don't really use e-mail," I cringed.

"That's how I let people know when the trainings are," she said agitated.

"Can you just give me the dates now?" I asked her.

"It would be easier if I could just e-mail you," she insisted.

My, my, so pushy! This was the first time I had encountered someone that was really pushing the e-mail issue. Probably because we had never met.

"Is there someone else in my daughter's class that is also doing clay?" I said.

"Yes, Teresa," she told me.

"OK, I'll just get the info from her," I said and ended the conversation.

Later, I called Teresa and explained that I'd be happy to help with clay, but was wondering if she could keep me informed of the training dates.

"I got the handout from the training, scanned it and e-mailed it to you, but got a bounce back," she told me.

"Yeah, I'm not on e-mail much," I explained.

"OK, I'll just meet you after school and give you the handout," she offered politely. From then on, Teresa was my go to gal for clay projects.

Day 172: "I'm being considered for a promotion," Jeff called from his office.

"What?" I practically screamed through the phone. "Finally!"

"But, the thing is…it's in Singapore," he said, waiting for my response.

"Oh, no. I'm sorry," I said. "You told them no, right? What a bummer. Well, it's still a compliment to be considered."

"Actually, I have until Monday to let them know if I'm interested. You know, it is a first world country," he said. "And it's surrounded by some amazing places that we could travel to."

"Why couldn't it be Italy or Spain," I whined. "I'd go in a heartbeat."

"Let's talk about it when I get home," he said.

Singapore. Never thought about living in Asia. Not my first choice because the culture is so vastly different and I don't know how I would learn the language. That night our conversation continued after Katelyn was in bed. Jeff, having traveled to Singapore more than once for business, was familiar with the country to some extent.

"It's very clean," he said trying to convince me to consider the option. "It's right on he water with a tropical climate. It's close to places like Bangkok and Bali."

"Is it safe?" I asked.

"I think so," he said. "The government is a dictatorship, so we'd have to see if there are any restrictions to travel."

"I just don't know anything about it. What language do they speak? What's the exchange rate? I wish we could look all this up online," I frowned.

"Don't we have an atlas?" said Jeff.

"Yes. I think we also need to invest in a set of encyclopedias," I joked.

As we flipped to the Singapore section of the atlas, we see a dot marking the country.

"Yeah, it's really small," said Jeff.

"This is crazy," I said, my eyes scanning the map. "I can't believe we are even entertaining the idea. It would definitely shake things up.

Quite a change from sitting in a corporate American cubicle for 10 hours a day. I think that living without technology has taught us to be more open-minded to opportunities. My only concern is that Katelyn would miss her cousins."

"It is a long haul, about a 20-hour flight," said Jeff. "Maybe you guys could spend the summer in Laguna Beach?"

"Without you? No. How much travel would you do? I don't want to move to a foreign country and spend less time with you."

"Most of my work would probably be in China and India, so it might mean a lot of travel, although I could go for a day or two."

"Doesn't seem like the best fit for our family," I said.

We decided to pass on the opportunity.

Day 174: As I rolled over on the Aerobed, I blinked my eyes trying to ascertain where I was. Oh yeah, Katelyn's room. My father-in-law and mother-in-law had hunkered down in our bed for the weekend while paying us a visit. After a breakfast feast of pancakes with lots of syrup, bacon and eggs, we were off to enjoy a day of playing tour guide. But first, feed and water the kittens. Upon opening the door to the office, I was besieged with the sight of diarrhea strewn across the floor -- making a trail to the litter box. *Great.* I hurriedly cleaned the mess and disinfected the area -- wishing I would've had industrial strength materials -- fearful the stench would permanently seep into the serenity of my office.

As we were about to leave, I reentered the office to leave some fresh food…once again I walked in on a fresh crime scene, unsure of the diarrhea culprit. I dialed the Humane Society to inform them that both kitties were now sick.

"It's probably worms," the woman informed me.

That's it! "I'm bringing them back," I said. "They've been sick since we got them. It's too much."

We loaded up the kitties and piled into the car to make the drop off. Luckily, Katelyn hadn't gotten attached. So much for our fostering experience. All I have to say is thank God it wasn't for six weeks! Later, I was talking to Bree who told me she fostered three puppies who tore up her house and she took them back early too.

Month Seven: You've Got No Mail

"The great myth of our times is that technology is communication."
--Libby Larsen, Grammy award-winning composer

Day 181: Katelyn picked up a hammer and, with my guidance, pounded a nail into a piece of wood. Her tiny arms swinging with surprising accuracy as I kept my hand far from the range of impact.

"This is fun, Mama," she beamed, as Bree and I looked on. "I've never used a real hammer."

We had made the trek to Berkeley to spend a technology-less morning at the Berkeley Adventure Park. Run by the city, the park provides kids with unstructured outdoor creativity.

Upon checking in, kids need to find and turn in debris in order to check out the hammer, nails, handsaws and paint. With more of the appearance of a junkyard rather than a park, children are encouraged to build and construct whatever their imaginations can conjure up, from forts to art. Various shapes and sizes of wood are scattered throughout, for use. Add a ramp to an existing structure, paint it pink, and presto-chango you've contributed to the collaborative effort to keep the park constantly changing. It's a continual work in progress, like a living art exhibit.

Katelyn and I chose to create "faces" from wood and paint. We then nailed them to some forts for beautification. In addition to building and playing, kids can also enjoy a zip line, slides, ropes, an old boat, a bucket drum set and more. The world needs more of these free parks where creativity can run wild along with the kids that build it.

In Menlo Park, Mike Lanza, started what he calls a "Playborhood," where he welcomes kids and families in his neighborhood to play together outdoors by doing a variety of interactive free play activities that he facilitates. His objective is to get kids to spend less time in front of screens.

According to a survey by the Kaiser Family Foundation and Center for Disease Control, a child is six times more likely to play a video game than ride a bike. With technology luring our kids indoors or organized sports dictating how our children's time is spent outdoors, we need more Playborhoods.

Day 183: Movie nights were something our family was missing. To compensate for the loss, I snuggled up with Katelyn, a fire roaring in the fireplace, a cozy blanket across our laps, as we listened to the story of Winnie the Pooh on our record player. The scritch-scratch of the record brought me back to my childhood. The record was something we had plucked from a thrift store while visiting Eureka recently. Unbeknownst to us, the record included an illustrated book of the story. We followed along in the book as the Pooh characters came to life through the speakers. We munched on popcorn and giggled at Tigger's mishaps. Suddenly, we were reenacting the same feelings as "movie night," but it felt better -- more interactive, imaginative, and personal.

Day 185: As of today, we have not watched TV at home for six months! Katelyn's still complaining. "When are we getting the TV back?" has become her mantra. Now it was time to be prepared for what the next six months without technology was going to look like.

Jeff said, "How's Katelyn going to watch the holiday TV shows?"

"At the cousins, when we're down for Thanksgiving," I said. "We can bring them with us."

"No, we're bringing back the TV for the holidays," Jeff insisted. "She's only young once."

His comment and retaliation hit hard. *Had he not learned anything from our six months without technology? Was it all in vain?*

"It's a tradition," he said. "I don't want her to miss it."

"She can watch them at a friend's house. Or we can go away for a weekend to a hotel with a TV. She won't miss out on them," I said.

***

"I'm cracking without technology," Jeff said. "I hate not being able to use my cell phone. I can't use my work phone to make personal call because there is no privacy. I used to use my commute time to make calls from the car. It kills me when my mom calls on my cell phone and I can't answer it. I'm losing touch with people -- personally and professionally. Career-wise I've lost a lot of contacts, it's a risk. I should be networking. I hate not knowing what's being sent to me on e-mail. What if headhunters are trying to reach me? It's not practical to go to the library to check my e-mail. I don't have time. My workload is too immense. We need Internet access at home. Last night broke me -- having to go back out at 9 p.m. and sit in a dark parking lot to get access to send my work e-mails."

Like an erupting volcano, Jeff finally exploded, letting his feelings and frustrations pour out like hot lava. Adjustments needed to be made -- for the health and well being of our family.

I cried, "I'm sorry I've put you and Katelyn through this. Maybe we should just give up. It's too much."

Seeing my pain, Jeff relented and we agreed we should continue, but with some modifications.

"First thing, is to reconnect Internet at home so that you can get your work done here without having to go out," I offered.

"That would be a huge relief," Jeff sighed.

"Why don't we have a date on Friday to discuss changes and implement them on Sunday -- our six month mark," I said.

We went to bed feeling better at having cleared the air.

Day 187: Why is it that moms tend to over commit themselves? The PTA, the soccer coach, the fundraiser chairman, Supermom. There are plenty of books that encourage us to "just say no" when asked to give of our time. I have been one of these moms -- always involved. Through giving up technology, I've been forced to take a break from volunteering as much as I used to. Without e-mail, Internet or Evites, I'm just not capable to step up. At first this saddened me. I've always enjoyed being involved. Now, I'm learning that rather than spending every free moment helping others, it's vital to take some of that time to myself -- reading a book, taking a class, attending a lecture, learning a skill.

As moms, we get so wrapped up in our kids' lives that we leave little time for ourselves. We feel compelled and obligated to say "yes" when asked if we can help at school, at Girl Scouts, at bible study. I've found an easy way out. Rather than saying "no" -- say "no e-mail." Use that as your excuse, "I'd love to be the room mom, but I'm not on e-mail right now, so unfortunately I won't be able to." By limiting your ability to help others, you'll be helping yourself, which is the best way to be a Super Mom.

Day 193: As Katelyn spun around the monkey bars, I surveyed the school playground in search of Sage. I was on a mission…in search of Charlie Brown. I spotted Sage and honed in on her.

"Hey Sage, I have a question for you," I said. "Do you know when the Halloween Charlie Brown special is on TV this week?"

"Not sure," she replied. "We got it from Netflix but just returned it. You really need to just dust off the TV. Just use it for movies only. Katelyn has learned how to live without it. She's already figured out other ways to entertain herself. She probably won't even want to really watch it."

"I'm afraid if we bring it back, she *will* want to watch it. And I don't want to battle with her about how much to watch," I explained.

"Sometimes not having something makes kids want it more. Just bring it back, but hide it in a TV armoire so she doesn't see it."

"I'll think about it," Ugh. *Six more months of defending my technology-less lifestyle.* So, whose house can I squat in so Katelyn can see the Great Pumpkin? I'd mull it over. Meanwhile, I had shopping to do.

As I browsed the Costco aisles, stacked from floor to ceiling with jumbo-sized boxes of Cheerios, refried beans, and canned soup, my eyes fell on the DVD section. And what to my wondering eyes should appear…the Great Pumpkin, Charlie Brown (plus Thanksgiving and Christmas with old blockhead, too). It was a prodigal purchase, seeing as how we didn't have a TV to watch them on, but at least this way our chance to see them increased since trying to watch them on a live TV at someone else's house might not pan out due to schedules.

As fate would have it, we were scheduled for a play date at Sandy's. Upon arriving, I unveiled my treasure. The girls squealed and begged to watch it. Sage granted their request. They squished into a rocker and lavished at the joy of a Halloween tradition.

"This show was on when I was a little girl," I told Katelyn. Nice to share traditions, even TV ones.

***

As our Prius glided silently home, Katelyn piped up from the back seat, "Mama, you know that old game thing we have that is kind of like a DS that I got to play with while driving down to see the cousins?" she asks.

"Yeah, it's called a Gameboy," I tell her.

"Well, can I play with that when we get home?"

Damn. I thought I had avoided the video game dilemma. *How in the heck did she pull the distant memory of that one car trip two years ago from her still-developing brain?*

I fib, "I'm not really sure where it is?"

"Oh, I know where it's at…in the garage, by our camping stuff."

How could I forget, my daughter's memory would put an elephant to shame, so of course she would know the exact spot that I shoved it.

I shuffle some boxes in the garage, attempting to appear as though I'm searching for the much sought-after toy. I just can't man up and tell her no. She persists.

"I know it's here, Mama. Just keep looking."

Instead I stumble across a stash of birthday presents I had purchased on sale to dole out to her friends throughout the year. I pick up a bracelet kit and suggest, "How about you play with this instead? After all, you've been so good about not having the TV, you deserve a present," I say showing her the kit.

"Oh yeah, yeah, that would be fun…but still keep looking for that game thingy."

The funny thing is, I don't know why we still have that Gameboy. I could probably sell it as an antique (if I had access to eBay). I purchased it for $200, as a gift for Jeff in 2000 as entertainment for the 14-hour Sydney to San Francisco flights we endured as ex-pats living in Australia. He never used it. It sat packed away in the garage for many years, until one day I happened upon it while looking for something else. After witnessing many (usually boys) youngsters glued to their portable gaming devices -- at the market, at restaurants, at the doctor's office -- we chose not to purchase Katelyn such a device. Kids should be aware of and appreciative of their environment, rather than living in a virtual Sims environment as a substitute. Although, these days, kids don't even need a DS -- moms willingly hand over their iPhones with games they've preloaded. I was guilty of whipping out my iPhone to intercept a tantrum on more than one occasion. The giggling baby on YouTube worked like a charm.

Day 194: Like usual, Jeff's tapping away on his laptop on the couch. I'm sitting across the room, reading a book. I pause and look up at him.

"Do you find it easier to get your work done without the TV on?" I ask.

"Heck yeah," he responds instantly. "I really don't think I'll go back to watching TV... just movies...and news.,,maybe sports...and I'd like to watch educational shows. Not crap TV. Well, maybe just one trashy show per week."

I start laughing. "So, you're not going to watch TV...only movies, news, sports, educational and occasionally crap."

With a scrunchy brow and a huff, he goes back to work.

"I'm telling ya, once it's back, it'll be hard not to watch it," I warn him.

***

Katelyn and her schoolmate Faith danced around the paved driveway sweeping brown, fallen leaves into a pile. The girls were jovial about what adults would consider a mundane task. *Swish, swish, giggle, giggle.* The rhythm of their play looking like the choreographed chimney sweep dancers from *Mary Poppins.* After a few moments, Faith asked, "Can we go inside and play with your market (that had replaced our TV)?"

"No, I want to do this so we can help the environment," Katelyn said.

"What's the environment?" Faith asked.

"It's the planet earth. Don't you know that word?"

Katelyn glanced up at me sitting on the porch, "Mama, did you know that old leaves are good for our garden?"

"Yes, but were did *you* learn that?" I ask.

"I read it in a book," Katelyn says nonchalantly.

My little budding environmentalist. I reveled in the idea that we were both learning something new together -- how to be greener.

***

Today marked six months of living without technology. To celebrate, we decided that we could have a 24-hour break in our techless life. I plugged in my laptop that I had hauled out of the garage. Katelyn scolded me, "Mommy, you're not supposed to be on the computer."

Its cold keyboard felt good beneath my fingers -- familiar, like a pair of worn jeans retrieved from summer storage. Today we were once again connected to the Internet -- at home. I hunkered down and got right to work, making the most of my fleeting, precious time with technology. First up, e-mail. I opened my personal e-mail and found 129 unread e-mails. *That's it?* As I scanned through them, I decided to unsubscribe from those auto listings that I felt no longer necessary -- those that I had taken the time and energy to sign up for and, at one point, seemed significant. Now, they appeared as SPAM -- a waste of my time. Do I really need continual e-mails from Costco, Safeway, several parenting websites, a local toy store, parks and recreation, travel sites? When my de-spamming was finished 45 minutes later, I had purged my inbox by unsubscribing from 38 newsletters. No wonder I was wasting so much time on e-mail!

My priorities have certainly changed. What once appeared as information and innocent entertainment, suddenly made way for a more significant use of my time. I had lost so many hours of my life reading about Safeway sale items, perusing travel listings of place that I'd never see, or learning about how to get my child to share her toys -- when I could have been playing tag with my daughter, reading an enthralling book, or making out with my husband. Clearing my e-mail inbox also lead to me logging off from several Yahoo Groups. I dismissed myself from 12 groups that I once found fundamental.

We often think of decluttering our home, but also a good practice to scale back on electronic communication. Commit to an annual digital spring-cleaning. After all, do you really need 200 bookmarks on your web browser, 25 files on your virtual desktop, newsletters on parenting now that your child is in college?

Digital photos are another huge time sync. It's your son's first day of kindergarten, you snap away knowing that your 2 GB memory card will allow you to capture the event in its entirety. Your trigger finger happily continues pressing the button. Upon arriving home, you upload your 623 images to your laptop (more on the card). One of two things happens, you either are overwhelmed by the thought of organizing that many photos and walk away from your computer vowing to "get to it later" – even though you never do. Your images, good and bad, make themselves comfortable knowing they'll be there a while. They won't be deleted, e-mailed or printed any time soon, if ever. Or, you decided to dig in and make the most of your artistic imagery, clicking through each one to discriminate which ones make the cut and which end up in the trash. Once narrowed down, you go through the "keepers" and digitally enhance them so as to remove blemishes and wrinkles, pump up the color, and increase the sharpness -- or completely erase an ex that ruined a perfectly good image of you and your friends.

Once that's done, you upload your favs to your Facebook, MySpace and Twitter. You transfer them to Shutterfly and take the time to painstakingly create a layout and design for a photo book to be printed and mailed to the grandparents. You can't forget to update your family blog with the photos and a complete caption describing what each one is. You might choose a couple stellar images and submit them online to photo contests looking for "the cutest kid." Lastly, you create a calendar to hang on your fridge next year so you can see how much your baby has grown. In the end, those 30 minutes of capturing a pivotal moment, steals at least eight hours of your precious time. The solution to this madness…be more choosey about what and how many images you take, and delete as you go. These simple steps will minimize your postproduction. Just because your camera is capable of taking 800 shots, doesn't mean you should.

As much as I am frustrated by the limitations of film, the simplicity is refreshing. I snap 36 images, get them developed then stick one copy in an album and one copy in the mail to relatives and I'm done.

Next, I check my work e-mail. Nothing. Katelyn is curled up with me, squished into an armchair in her jammies, reading a book.

We decided not to lug the TV out, but I did inform her that she could watch a movie on my laptop once I finished my work. With Jeff off enjoying a mountain bike ride, Katelyn eventually lost her patience of entertaining herself and found it useful to continually inquire, "Are you done yet, Mama? Can I watch my movie now?"

I managed to stave her off while I tried to quickly and efficiently complete my technology tasks. Next up, informing my photography clients (right before my normally busiest time of the year) that I would not be shooting their annual holiday card family photos this year, or ever. It was a letter I sent with hesitation but felt obligated to close that chapter for good. Upon hitting send, I felt a bit of sorrow remembering the beautiful people I had met through my portrait photography business. It has been a joy watching their families grow from pregnancy through kindergarten. The past six years as a photographer had allowed me to tell their stories through the lense of a camera. And it had allowed me the flexibility to spend more time with my own family. Hard to walk away from a rewarding career, but time for the next chapter.

Handing over my laptop to Katelyn, she positioned it on her lap and quickly became engrossed in *Barbie Dear Diary*.

Jeff's laptop sat unused on the kitchen table. I opened it and continued my tech frenzy. There were now a whopping 14 people following me on Twitter, even though I had never sent a tweet. I perused

Facebook, catching up on the trivial goings-on in my virtual world of friendship. After having tried to track down a couple of old friends without luck, I did find a message from one of my very best friends that I had completely lost touch with since unplugging -- despite my continual phone calls to her. Her message informed me that she only communicates through text. So apparently, it seemed as though she would not be a part of my life since I was not texting. Sad, but true. A clear sign that society has become more comfortable with communicating through technology, at the detriment to personal relationships.

Nick Andes and Doug Klinger took texting to the extreme, trying to apparently set a record by sending a total of 217,000 text messages in one month -- about 7,000 a day. Luckily, after realizing they didn't have "unlimited texting," T-Mobile refunded their bill of more than $26,000.

Face-to-face or even voice-to-voice conversations have nearly become archaic. I fear that the creators of *Wall-e* had an eerie prediction when they depicted humans utterly dependent of socializing through computers. *Is that truly our fate? Becoming obese and disconnected?* If something doesn't change, that could very well be the direction humanity takes.

Next, shopping online. First, a custom sterling silver necklace with "Jeff" engraved on the front of a small circle pendant and our wedding date engraved on the back. A constant reminder of how lucky I am to have him as my husband.

On to Craigslist…my long lost friend. In search of an elliptical and swing set. I browsed several listings hoping to hit the jackpot. They're either too big, too expensive or too far away. A swing set to lure Katelyn into the backyard more often, and to selfishly keep play dates from causing my home to appear as though the Tasmanian devil stopped by. And an elliptical to sneak in my rainy day workouts since the gym is becoming less and less attractive with the lure of the TVs. Once on Craigslist, my mind automatically starts to convince me that there might be even more things I need to buy, after all…they are cheap, why not? How about a porch swing, a fire pit, a new dining room table. I notice an internal conflict brewing, buying used is better for the planet, but might not be as good for my wallet as I think.

I am always a sucker for "sale" items. While in line at Target last week, an older woman in front of me said to the cashier, "I always come in here to buy a couple things, and leave with more than I intended." The cashier replied, "That's our intention." *OK, that's it. No more shopping at Target.*

In the parking lot, I noticed the same woman wandering aimlessly, pressing the panic button on her keychain to sound her car alarm. As it beeped loudly throughout the lot, the woman followed the beeping in order to find her car. That was me, a mere six months ago. So busy worrying about what I had to buy or do, that I never bothered living in the moment and pay attention to where I parked. In college, I had parked in a multi-level lot and was convinced my car was stolen after walking several levels in search of it. I informed the gate agent, who zipped me around in his cart, eventually landing at my car, right where I had left it. Living without technology brings clarity of mind.

After shutting down our laptops, I grabbed the fully charged video camera that had been sitting untouched for six months. I turned it on and pointed it at Katelyn. The occasionally camera shy six-year-old soaked up the spotlight and hammed it up for the camera. I questioned her on tape about our no-technology lifestyle. She voiced her complaints with enthusiasm. Then I turned the camera over to her, somewhat chuckling at her lack of experience in using the camera. Had we not unplugged, I'm completely confident that not only would she currently be proficient at filming, but I'm sure by this age she would also know how to edit in iMovie and upload to YouTube.

She positioned the camera in my general direction, her small hand grasping it tightly. As she jiggled the camera side to side and up and down, attempting to center me in the frame, she giggled zealously. She then took on the task of interviewer and questioned me with authority. Lastly, I taped Jeff, on his laptop. He had nothing nice to say about doing without. He was still fuming about the discovery that AOL only saves unread e-mail for 30 days. Since he had not checked it in six months, there were hundreds of e-mails sent that he never received. "It's like waking up on Christmas morning with no presents," he said.

With half our day sucked up by technology, we piled into the car and headed downtown. Katelyn and Jeff grabbed some pizza while I went for a run…with my iPod!

While jogging along the Los Gatos Creek Trail, I had Madonna, Ella Fitzgerald, Britney Spears, Barry White and Edie Brickell to keep me company. I welcomed them with open arms. They helped me pick up my pace. I turned up the volume and sang out loud, "Womanizer, womanizer, womanizer, womanizer." Hopefully people passing by wouldn't think I was badmouthing my husband. Heaven knows what they thought when I belted out, "I kissed a girl, and I liked it." When I met up with Jeff and Katelyn, sweaty and exhilarated, I said, "I could've run for an hour. I had forgotten how much easier it is with music."

Our technology indulgence ended with Katelyn and I cozied up on he couch watching *Charlie Brown Thanksgiving* while Jeff shopped for bike parts on eBay.

Day 201: With cool autumn weather upon us, plants at Hidden Villa beckoned to be moved to the warmth of the green house. As we tidied up the greenhouse, blue-bellied lizards scattered to and fro, unsure about the ruckus from the intruders. My fellow volunteers and I swept the dirt-covered cement floor, pulled weeds, and discarded crumbling plastic pots that had been forgotten about since last winter. My hands encased in durable leather gardening gloves, I bent down to gather a pile of pulled weeds. A lazy lizard sprang from the disturbed pile and I let out a yelp and dropped the dead foliage. My body shook with the "heebie-jeebies."

"I love to look at lizards," I explained to my chuckling colleagues, "but I have no interest in touching one. Yuck!"

After we finished spring-cleaning the greenhouse, we moved to the educational garden to plant some Jerusalem Sage and as I bent down to dig a hole, I felt a poke in my side. I surveyed the surroundings for a thorny bush. None. I continued working and felt the twinge again and then stood up and it went away. Maybe a strained muscle? Once more, as I knelt in the dirt, there was a twitch in my side. As I rose to my feet, I pulled my elastic waistband open and stuck my hand down my pants to see if I could feel anything. I screamed as I pulled a blue-bellied lizard from my pants and flung it high into the air. "Ew, ew, ew!" I wailed as I shook my arms and jumped around in disgust. My fellow gardener, Diane, a gentle middle-aged woman took sympathy on me and did not bust out in laughter, but asked if I was OK.

"Do lizards bite?" I questioned, easing my pants past my hip where the lizard had chosen to nuzzle. No bite marks. Although the garden was full of school groups touring the farm, the children hadn't noticed my antics, thankfully.

With a vow not to wear baggy-legged pants for gardening in the future (the lizard must have crawled up my pant leg), Diane and I continued with our planting. We had only worked together a couple of times, but I was always impressed with her depth of knowledge about plants.

"Do you have a garden at your home?" I asked as I jabbed the shovel into the moist earth.

"Not really," she said as she removed her sun hat to wipe the sweat from her brow. "But we do grow grapes at our home in Walnut Creek."

"So, do you make wine?"

"Yes, my husband has been doing it for 10 years."

I stopped and put my soiled hand on my hip and said in dismay, "That is so weird. I was just having dinner with a friend last night and I mentioned to her that I've wanted to learn how to make wine. Does your husband need an apprentice?"

"As a matter of fact, we are having a wine bottling party this weekend. Why don't you join us," she said.

"Really? I'd love too!"

"Great. Then my husband can show you the equipment he uses and the process it takes."

Leaving Hidden Villa, I was in awe of the power of the universe. Something that had recently been a wish, was being granted by a higher power.

Day 203: Arriving 45 minutes late, I knocked on Diane's door. She answered and greeted me with a hug. We meandered our way through her home and out to the back patio where a group was gathered around a large table sprinkled with appetizers. With introductions all around, I came to learn that I was privileged to be sitting with the board of the Mid-Peninsula Open Space, an organization that preserved our natural land and its beauty. In admiration, I made a point to converse with as many as I could.

After small talk, it was time to get to work. Diane gave me a tour of their wine cellar and workroom, then her husband gathered us into a group and explained the bottling process. First, the wine would flow from a large vat to an individual bottle. Then the oxygen was removed and a cork secured. Lastly a label was applied. We rotated jobs as we sipped wine. A cheerful, older crowd -- they worked diligently and merrily. Our assembly line yielded 50 bottles of four-year-old Port. I had tasted new wine, made new friends and learned about a potential new hobby. Upon leaving the party, bottle of Port in hand, I had hoped to continue to assist in future winemaking efforts.

Day 204: Pulling into the parking lot, the vanity plates on the cars foreshadowed the day that was about to unfold, "YogaPry" and "ZenLimo."

As my friend Bree and I stepped out of her car, the air bit my nose. We strolled along the gravel path towards the Green Gulch Farm Zen Center in Marin County -- a 115-acre Zen Buddhist practice center and farm founded in 1972.

While indulging in a chanterelle, Portobello and crimini mushroom omelet at Greens vegetarian restaurant in San Francisco recently, I had learned that it was established by the San Francisco Zen Center in 1979 and that Green Gulch provides food from its organic garden to the restaurant. I just had to see what a monk-maintained organic farm looked and felt like.

Upon investigation, I learned of a Sunday lunch and meditation that was open to the public for only ten dollars. The program included zazen instruction, a Dharma talk, and tea. Bree was up for the adventure, too.

A group gathered outside the Zendo (temple), their breath visible in the cold air. We joined them and heeded the sign, "Please remove your shoes" – thankful that I had worn warm socks. A female monk greeted us, dressed in a gold and maroon robe. As she bowed, I could see the top of her closely shaven head. As we were escorted into the Zendo, we were instructed to bow upon entering.

Inside were neat rows of round cushions that reminded me of a checkerboard. We plunked ourselves down on a cushion and tried to adjust to a comfortable sitting position. Our meditation instruction began. "Stand. Palms together. Elbows bent at the waist. Bow to your pillow to thank yourself. Bow away from your pillow to thank the others," said the monk guiding us. As Bree and I bowed awkwardly, we tried to stifle our giggles.

The incense tickled my nose as the monk ran through various sitting styles -- cross-legged, lotus, kneeling. Our hands were placed against our abdomen with our thumbs and fingers creating a circle.

Next, breathing. "Focus on the breath -- moving slowly in and out," he said in a calm tone of voice. "Breathe from your abdomen." *Sitting -- check. Hands -- check. Breathing -- check. Eyes -- half closed.* I struggled to keep my eyelids at half-mast. They fluttered, wanting to open or close completely. My squinted eyes blocked out most of the light, but I could still make out shadows. Maybe the eyes half-closed was a liability issue. After all, there were close to 200 people filling the room. What if people started dozing off and keeling over, bumping into the person next to them. The meditation could turn into mayhem.

"Let your thoughts enter. Acknowledge them and then watch them float past like a cloud. Count your breath to 10, then start over," the monk continued.

Once instruction was complete we did a practice run, a two-minute meditation. I adjusted my rump on the cushion, crossed my legs, lowered my eyes…and sat. The crisp air penetrated through the walls and chilled me. This was nothing like bikram yoga with its 102-degree room. *Think warm thoughts. Think warm thoughts.* A chime softly rang, declaring the end of the session. Next, it was time to get serious, 40 minutes. *Could I do it?* We stood, stretched, bowed and resettled. Ready, set, meditate. I tried to sit quietly and calm my being, but suddenly a wave of panic washed over me. Like a caged lion, could I sit "restrained" for such a long period? I had a flashback to my root canal, stuck in the dentist's chair, drill in my mouth, waiting for the torture to end.

I felt an urge to run screaming from the room, but I resisted. *Breath -- one, two, three…*I snuck a peek at Bree. She must have sensed my movement because she raised her eyelids slightly to glance back at me. I quickly faced forward in fear of being scolded by the monk. Something so obviously serious somehow seemed silly and I swallowed my laughter. Since high school, Bree and I always know how to have a good laugh together -- whatever the situation.

OK, breeeath. In. Out. What's all that racket outside the door? Someone coughing. Someone chatting. Someone's heels clicking on the wooden planks -- oh, they forgot to take off their shoes! Breath. Let it pass.

My nose itches. Am I allowed to scratch it? Oh, it really itches. Scratch, scratch, ah. Breath. Now my knee itches. Crap. A quick scratch. Reposition my hands. *One -- breathe in. Two -- breathe out. Damn. It's cold in here. Ching…*

"Slowly wiggle your fingers. Open your eyes," the soft voice instructed us.

What? There's no way that was 40 minutes. It felt like five. Glancing at my watch it felt surreal, like I briefly entered a world without time. No past, no future -- only the present.

I felt a bit jarred upon re-entering real life, like a bear leaving its warm cave after hibernation. While meditating, although distracted, my senses seemed muted, softened. Afterwards, noises seemed louder, scents stronger, my body felt heavy. After bowing to our pillows, we filed out into the yard and were treated to steaming hot green tea and monk-made apple muffins. Tables lined the lawn selling fresh produce from the garden, a mini-farmer's market.

Bree and I stretched our lotus legs with a walk through the immense garden and farm. Not as picturesque as Hidden Villa's educational garden, this was a real working farm. No cutesy hand-painted markers to identify the crops. No ivy tunnels to crawl through. No sunflower canopies. This was a flourishing, rudimentary plot used to produce nourishment.

After tea, we headed indoors and were treated to a lecture by one of the head monks that was being filmed for a show that would air on PBS, which I wouldn't see since I was living unplugged.

After some pomp and circumstance, and quite a few minutes adjusting his cumbersome robe, he welcomed the crowd, perched on his cushion. As he indulged us with his meaning of Zen, my mind wandered, but was pulled back by the word, "Google."

"Where you always get a good lunch," he laughed. He talked about training engineers at Google to meditate. "I call it SIY – Search Inside Yourself," he said. "Engineers always want to fix and solve problems. They need to learn to just sit."

Here we were in a serene setting and technology had seeped its way in.

I left the lecture with two kernels of wisdom that I hoped would help my life expand and grow like the garden I had wandered though -- one, just because it doesn't interest you, doesn't mean you shouldn't learn about it. Two, when encountering something negative in life, you need to realize that it is what it is -- and let it go.

As we slowly and quietly slipped our shoes back on and trotted to the lunchroom, Bree and I shared our thoughts over a bowl of fresh Lentil soup and garden-grown salad.

"Thanks for joining me today," I said to Bree, while blowing on my hot soup.

"Yeah, this was quite an experience. It's always an adventure with you," she laughed.

"You know, they have a seven-day meditation retreat. You start your day at 4 a.m. You in?" I winked, half jokingly.

"Ah, no. I'll leave that one for you," she said shaking her head.

As we went our separate ways home, I left feeling grateful for having such a true friend, the courage to explore alternative ways of living, and a thirst for adventure.

Day 205: With a day off school, I thought we would do something educational and entertaining…a visit to the Computer History Museum in Mountain View. Katelyn wasn't so sure she wanted to waste her precious time off of school looking at a bunch of old computers. Well, that's putting it mildly. She actually protested loudly. She might as well have been holding a picket sign that said, "Hell no, I won't go." But we went.

Upon entering the museum, we made our way to the room full of relics. There, on a cold, metal shelf, surrounded by dozens of prehistoric computers, sat the original Apple personal computer, the Macintosh II.

I snapped a photo with my disposable camera.

Katelyn rolled her eyes.

I was instantly transported back to 1983 -- Winship Junior High school. A boxy beige computer sat at the back of the room -- its slot waiting to swallow a floppy disc. Winship's "Viking Saga" newsletter published an announcement of new computers, "Even at Winship, the computer craze is catching on. We recently purchased two new TRS-80 Model 3 microcomputers from Radio Shack. They both have single disc drives and a printer. A disc drive plays these little discs that hold information about programs and games. The computers are used by teachers and GATE students now, but next year more students will be able to use them."

As we circled the room going further back in time, we came to a photo of Steve Jobs and Steve Wozniak, founders of Apple. The photo was taken in 1976 when they made and sold their first version of a ready-to-assemble computer, Apple I. The actual product sat next to the photo, encased in plastic. They sold it for $666.66. Like many technology giants in the area, were they aware of what their humble beginnings would develop into?

Continuing our journey through the museum, we came upon an Atari *Pong* game. When I was a child, in the early '70s, we had one in our "game room" in the attic. Bouncing a digital ball across a screen -- it didn't get more exciting then that...until *Pac-Man* came along.

We left the museum with a robot...a ladybug robot, that Katelyn had to have from the gift shop.

***

Pulling weeds at Hidden Villa, I told Dorothy about my excursion to the Computer History Museum. Dorothy, who is quite a few years old than me, stated casually, "Oh yeah, I had dinner with Steve Jobs one time. It must have been in 1980."

Suddenly she became a rock star in my eyes. "What? How many people in Silicon Valley can say they've chatted over a meal with the inventor of Apple!" I said to her. "I remember when I was working as a web producer at ZDTV and Microsoft's Paul Allen came to give a lecture. The higher-ups were bouncing off the roof. They were so excited. Apparently the CEO and VP went to his house in Seattle and said it was over the top."

"Yeah, I was in Seattle when Bill Gates was building his house and boy were the neighbors not happy with the immense construction," Dorothy said. "It's funny because when I was young, I knew a few of the guys who were just starting to break into the technology industry. Now I go to the Computer History Museum and see their photos hanging on the walls."

My father-in-law recently recommended that I read the book, *The 100 Most Influential People in History*. Certainly Bill Gates and Steve Jobs should rank up there with Shakespeare, Hitler, Thomas Edison, Elvis, and Jesus.

***

I'm beginning to feel like the captain of The Bounty. Any day now there's going to be a mutiny. Jeff and Katelyn are going to make me walk the plank so they can return the TV to its rightful place in the living room. Better watch my back.

Month Eight: Wii Wish You a Merry Christmas

"TV and the Internet are good because they keep stupid people from spending too much time out in public."
--Douglas Coupland, Canadian novelist and artist

Day 215: While shopping for some organic ham-flavored turkey at Whole Foods, Katelyn spouted, "I have to go potty, Mama."

I whisked her to the toilets, but it was occupied. As we waited I toiled over a community bulletin board and saw an ad that read, "Family Meditation." I plucked the flyer and shoved it in my purse.

At home, while Katelyn gobbled up her turkey-ham, I unfolded the flyer. "Meditation for Families with kids age 6 and up." *Hmm. Meditation for kids.* Something I hadn't thought of. *Can kids really sit still and contemplate their being?* This I had to see.

"Katelyn, want to learn how to meditate?" I asked her as she sat next to me at the kitchen table.

"What's that?" she said.

"Well, it's learning how to keep your body calm and relaxed," I explained.

"Sounds boring," she replied.

Do I admit to her that, yeah, it is boring, but that it would lead her to a deeper, more meaningful life? She could sit in silence and contemplate how to become teacher's pet, what it would take to get Mama to let her eat her advent calendar chocolate at breakfast, and why the boys like to chase her around the playground at recess for no apparent reason.

Day 217: Needing to be convinced myself that meditation was good for the whole family, I decided to leave the family behind and do some meditation reconnaissance. One episode of *Wife Swap* featured a family that meditated. At the time it aired, I was in full technology mode, living a plugged in life. It struck me as sad that the militant father head required his children to sit stoically for extended periods of time. *Shouldn't they be romping in the woods, indulging in video games or chatting endlessly on the phone?*

Just minutes from my home, I entered the Universal Unitarian church. Seated at a fold-up metal table in the entryway was a middle-aged woman monk. She greeted me warmly and collected my ten-dollar fee. The meditation students gathered in the chapel, awaiting instruction. As I shifted my body weight in the metal chair, a young boy and his father took a seat behind me. The boy fidgeted restlessly. I turned to the father and said, "How did you get him to come? My daughter is about his age and she wanted nothing to do with this."

The boy chimed in with an air of confidence and answered for himself, "Oh, I know what meditation is. I have a book about it…and Buddha." He appeared to be genuinely curious.

The three-hour session began with a children's story related to living a calm life, then the children were escorted to a different room for some hands-on meditation practice that involved sponge painting.

Accompanied by about a dozen adults, I entered my bliss more easily than last time, but couldn't help but wonder whether the kids' room looked like a scene from *The Sound of Music* or *Daddy Daycare*. After a brief intermission, I chose to spend the second half of the session with the kids.

The sunlight streamed into the room, warming the circle of round cushions as the children found a spot to rest. The half-dozen shoeless kids, ages five to eight, sat restlessly as their teacher searched for the words to describe an intangible process. As the teacher reviewed the morning meditation, she encouraged the kids to try again.

"Close your eyes. Sit still. Keep your hands to yourself. Breath," she said inhaling deeply and exhaling loudly.

During the course of the five-minute meditation session, the kids shifted like windblown sand. Most could not keep their eyes sealed shut, glancing curiously at their neighbors. The teacher gently guided them back to stillness. Two minutes in, Max, the boy I met earlier, announced to the assistant that he had to pee. She escorted him to the bathroom.

Upon returning, he tried to center himself but was distracted by his sticky nametag. Taking it off and on, off and on. The eldest girl sat like a statue. Her face drained of its childlike whimsy. Her legs crossed. Her hands resting on her knees, middle fingers and thumbs touching. She had obviously done this before. The boys were squirrely, needing to constantly keep their bodies in motion, even if it meant just wiggling toes.

I was amazed at the level of commitment these children possessed. They were devoted. They approached the task with maturity beyond their years. There were no giggles, no elbows jabbing, no goofing off. I had never witnessed a child sitting willingly with nothing to do.

"During the first session we learned that our mind is like a sponge," the instructor told the kids. "You can soak up the positive (the yellow paint) or the negative (the black paint). We should always strive to fill our mind with positive thoughts. Now we are going to learn how the mind is like a balloon, it likes to drift away in thought. You need to hold your mind steady and let bad thoughts float by like clouds. What are some bad thoughts we should let go of?"

A girl donning an Apple computer T-shirt raised her hand enthusiastically, "Noise, like when the phone rings."

Max added, "TV distracts me."

Another boy said, "Video games."

A girl piped up, "When my baby brother cries. It's annoying."

A boy added, "When my sister hits me."

Siblings and technology were earmarked for the cons column. Those things seemed to be weighing heavily on the minds of the young. The teacher's solution to the tug between good and evil is to be mindful. She doled out golden velvet bags that encased black and white stones.

"I keep this bag of stones with me to help me remember to be mindful of my thoughts and to try to think positively," she told the children. "One day, I was driving to work and complaining about the traffic."

She holds up a black stone.

"That's a negative thought, so I told myself to be thankful that I even had a car to drive to get to work."

She holds up a white stone.

"I changed it into a positive thought. When I got to school, the principal was angry with me for being late, which made me upset. Black rock. Then I thought that I didn't want to enter my classroom with a frown. I wanted to greet my kids with a smile. White stone."

The children seemed to cling to the idea like a mussel on a sea-drenched rock and clutched their own precious stones to take home. Wandering up to the teacher after the kids dispersed, I asked, "So, I have this neighbor with a relentless barking dog that drives me crazy. How do I let that go?"

"You could feel compassion for the dog since he's probably not getting the attention it needs or you could use the bark as a signal to be grateful for your life, a reminder that if a barking dog is the worst of your problems, you have a good life," she counseled me.

Yes, a barking dog was not a big worry.

At home, I handed Katelyn the bag of stones the teacher had given me. I recanted the story, unsure if it sunk in. Treasuring her gems, she scoured her room for the perfect box to hold them. She emerged holding a blue stone.

"Mama, remember this? Mrs. Nelsen gave it to me before school started. I wonder what 'blue' means?"

"Maybe it's a wishing stone. Or a gratitude stone," I say examining the stone.

Katelyn cocked her head, her eyes pointing toward the heavens in contemplation, "I know! It's a courage stone. When I'm afraid to do something, I can just grab my blue stone to make me brave."

"Perfect!" I smiled proudly at her. She gets it.

Day 218: At Trader Joe's, as the cashier stuffed our groceries into our reusable bags, Katelyn's friend Sandy, who had joined us, said, "Let's go get a balloon!"

Katelyn, glued to my hip since birth, was unsure about prying herself away from me and braving the five feet to the balloon counter to talk to the strange man that was handing out balloons.

"Come on, Katelyn," Sandy pulled at her.

"Go ahead, you'll be fine. I can see you from here," I encouraged.

Katelyn pulled my ear to her lips and whispered, "I need my blue stone."

I squatted down, looked her in the eyes and said, "The good thing about the stones is that you don't really need them with you. You can just think about them. Imagine that your blue stone is with you."

I kissed her on the head and gently pushed her towards the balloon counter. She returned victoriously with her new balloon and sense of confidence -- without the stone in her possession.

Day 221: The day before Thanksgiving…we were not buying turkey, potatoes or pumpkin pies. We were going to Disneyland! Having arrived at Marie's house the day prior, Katelyn bounded out of bed like Tigger ready to pounce on her cousins. The plan -- Jeff and I would take

Katelyn and the cousins, then Marie would meet us after her holiday grocery run.

As we prepared to head out the door, Marie made a discouraging discovery…their Disneyland passes had expired. Thinking quickly, she booted up her computer to purchase new ones from Disneyland.com. The five-minute project turned into 45 minutes with the difficulty navigating the website, retrieving a password and massaging a cranky printer. Finally, success. We flew out the door like Santa on his sleigh and made our way to the happiest place on earth.

Main Street was filled with throngs of mouse ear-wearing Disney fans. Lines were long, patience was short. I kept eyeing my watch to ensure we didn't miss our rendezvous with Marie -- eleven at the Mexican restaurant. Without our cell phones, missing Marie at our predetermined destination would be like a jet refueling in mid-air. You get one chance. If you miss it -- the plane will crash.

Navigating Disneyland with a group of five and limited cell phones forced us to stick together and communicate concisely. By late afternoon, with only a half-dozen rides under our belt, we came to the rationalization that the kingdom – with long lines and too many people -- had lost its magic, and headed home.

Holidays with the family, in the past, had meant a chaotic gift exchange with the cousins -- wrapping paper flying off the gifts in a frenzy. Gifts tossed aside before the to/from tag could be read. Cameras flashing. Kids squealing. And at the end of it all, a mound of gifts sat ignored as the kids moved on to the next activity, usually playing Wii, watching a movie or riding bikes.

This year Marie suggested we try something new. No gifts -- seemed like a responsible act in such a sour economy. Instead, the cousins gathered near a faux Christmas tree in the living room. Seated around a fold-up table, the kids were engrossed in decorating tree ornaments. The table was filled with glitter, glue, markers, ribbons and four glass balls per child. They chatted, giggled, sang, and bonded. Each leaving with creations they could proudly hang on their tree. Having switched the focus from getting to creating seemed to be the right way to celebrate the holidays. Creating ornaments, creating bonds, creating memories.

Day 222: The mashed potatoes were creamy, the cranberries sweet, the rolls warm. My spoon sunk easily into my meat-free stuffing-filled delicata squash, made for me with love by my mother-in-law. As four generations gave thanks for the meal before us, the home filled with warmth. Thanksgiving included delectable food, stimulating conversation, and the joy of being with family. The post entertainment? No football. A family-style, *American Idol* karaoke sing off on the Wii.

Katelyn sang softly into the microphone as Brandy pranced around the living room, shaking her hips and flinging her hair as the backup dancer. The rest of us were the screaming rock star fans. A group effort. One-by-one, young and old took the mic. Adults were treated to all five cousins being backup dancers. Thanks to the Wii, we were embarrassed, elated, and entertained.

In attendance at Thanksgiving was Jeff's last living grandparent, his grandmother, nicknamed "Mom." At age 90, she always had the most intriguing stories of life before the industrial revolution. Born in 1919, her childhood was simple. No Internet, no Wii, no cell phones, no television.

"We did own one of the first Ford cars," she recalled. "It didn't have seat belts and one time as we went around a corner, I fell out, but I wasn't hurt."

Over the school break, Katelyn had a Family History Project to complete. When I told Jeff's grandmother about it, she was thrilled to fill us in on the family's past and invited us to her home the next day.

With our family tree complete, she blessed us with an extensive genealogy research that had been done, leading back to Mayflower Captain Miles Standish. She dug out her mother's and father's eulogy that I read with fervor.

Her father was president of the Sierra Club. Her brother had been one of the first mountaineers to conquer many of Yosemite's peaks. And her sister spent time tagging along with Ansel Adams in the Sierra. Having recently become members of the Sierra Club, because of our love of the outdoors and desire to protect it, I hadn't realized the bloodline connection.

Day 224: At my father-in-law and mother-in-law's serene home in Laguna Beach's Bluebird canyon, we joked that we would pay for our night's stay by offering free technology services, something they were always in need of and we were happy to provide.

First on the list, I had to make up for an ignorant or lazy cable guy that had not programmed the remote to operate the DVD player. My father-in-law had spent the past few months saddled up to the TV while movie viewing so he could reach the pause and rewind buttons on the DVD player. As I read the manuals, checked the wiring and pressed some buttons -- voila! My father-in-law now had a fully functioning entertainment system.

Katelyn was treated to a Christmas movie as Jeff and I continued our good deed. Jeff helped his dad navigate Netflix while I assisted my mother-in-law with how to free the digital images captured on her camera. Jeff, done with his duty and bored, set out on a treasure hunt from his childhood.

"Mother, do you know where my old baseball cards are?" he asked.

She paused. "They might be here in the window box," she said, motioning to a box seat under the window near us.

Jeff pried the seat open like a time capsule. The items inside had been locked away, untouched for 20 years or more. My mother-in-law and I refocused on our task. Plug the camera into the laptop, upload images to iPhoto, label, organize, and create albums.

Jeff kept us informed of his uncovered treasures -- a size 5T well-worn wetsuit, some baby shoes, 1980s editions of *Surfer* magazine.

"How do I get actual prints made of my photos?" My mother-in-law asked. "Bring the camera to CVS?"

"No. You don't even need to leave the house," I told her. "You can order them online and have them mailed to you."

I could see the wonder in her face. She was giddy. A newly established Shutterfly account, a few clicks of the mouse and her prints would arrive in a couple days.

"Oh, I just got an e-mail confirming my order!" she said, checking her AOL account.

Just then Jeff interrupted, "I found them…my baseball cards!" He also uncovered several dusty textbooks and writing assignments from his days at Thurston Middle School. The juxtaposition of the day was memorable. On one side of the room, high-technology happenings. On the other side, no-tech relics. The timeline from 1970 to 2009 stretched from one end to the other.

Katelyn's eyes grew wide when Jeff presented her with his gem and mineral collection from the '70s. A true treasure!

Day 225: Time to deck the halls, starting with a tree. We decided to buck the system and forego purchasing our Christmas tree at Home Depot. Instead, we opted for a less capitalistic, more back-to-nature approach. A short drive into the Santa Cruz Mountains led us to several Christmas tree farms. At Four Winds, we were handed a small handsaw and told, "Pick any tree for fifty dollars."

Acres of pine tree-covered land lay before us. Wearing my Santa hat, I sang, "We wish you a Merry Christmas" as we weaved through the maze of trees. Katelyn had been reluctant to "cut down" rather than purchase it because, "If we cut down the trees, that's not good. We need trees. They help us breathe." I reassured her that new trees would be planted. Still, she hesitated. I breathed in the fresh sent and was reminded of my childhood. Every year we cut down our tree in Eureka. There was no Home Depot. Happy to share this tradition with Katelyn, we marched on in search of the perfect tree.

Too tall. Too bushy. Too scrawny. Too sparse. Like Goldilocks in search of the perfect bed, we found one that was "just right." Jeff bent down, grabbed the knotted trunk and began sawing. *Back and forth, back and forth* until it tipped. "Timber!" we yelled in unison. We hauled the tree back to the car feeling like real lumberjacks.

Trimming the tree took on a new twist. As we hung our hand-made ornaments on the prickly branches, our record player went round and round, spewing out classic Christmas carols from John Mathis and Burl Ives. This year we would not indulge in our surround sound digital holiday songs by our favorite new artists.

After the tree twinkled with lights and glass balls, we felt a bit letdown to do without our tradition of watching *The Grinch*. Without our TV, there would be many holiday specials we'd miss out on. With Katelyn growing older by the minute, we knew our time of sharing these specials together was limited. My heart sunk as I longed for our TV.

Day 226: Living a life with less focus on consumerism means having less garbage. Our measly trashcan was only half full this week. A sign that we were now less influenced by advertising and were buying less. We weren't ready to take it to the extremes that "No Impact Man" did, doing without his refrigerator, car, electricity, but our small effort was making a difference.

***

When I walk into a bookstore, I believe books choose me. Whatever I am meant to read at that time in my life will stand out. Six months ago, I picked up *Three cups of Tea*. It sat on my shelf waiting for the right time to be read. This month, I plucked it from the lineup and was instantly enthralled. The sacrifices Greg Mortenson made to help those less fortunate truly touched me. The fact that he was willing to live in poverty to keep his promise to the poor, made me realize that a lot of my purchases, that seemed necessary, were really frivolous.

Do I really need a new fleece coat at REI when I have a closet full of other coats and many people don't even own one. How can I cut back and start using some money for a good cause?

Having grown up in a capitalistic society, it's in our nature to want more. A better car, bigger house, fancier clothes. It's sad really, because our intrinsic desire for material objects robs us of the ability to live a truly satisfying and rewarding life. We are always left wanting more -- never satisfied with what we have.

Greg was sleeping on the floor of a hallway to save on rent so he could get back to Pakistan. Yet, even that hard life is nothing compared to what some people in third world countries endure -- mud homes, no toilets, no clean water, no electricity, little food or medical care. Even the poorest of the poor in the U.S. are living large compared to those half a world away.

And here I was spending five dollars on a Christmas tree ornament when those five dollars could literally change someone's life. Life's biggest gifts do not come from a store.

Day 233: The day after I finished reading *Three cups of Tea*, I walked into the bookstore and there, propped up on a table in the front of the store, was a book with a familiar face on the cover. Greg Mortenson's newly released book, *Stones into Schools.*

Later that night I told Jeff over dinner, "Isn't it weird that I had this book on my shelf for months and the day after I finish it, his new book is released and I just happen to go to the bookstore that day? You have to read this book."

As Jeff cleared away the dinner dishes, I unfolded the newspaper and low and behold stumbled upon an event listing…Greg Mortenson was giving a lecture the following Saturday at a college 15 minutes from or home. We had to go!

Day 235: The dark and rainy night, Katelyn's cough, Jeff's upset stomach -- could not keep us away from hearing Greg Mortensen tell his story. Upon arriving at Foothill College, there was a long line. As it turned out, the event, which garnered thirty dollars a seat, was sold out. Folks waited in the drizzle -- hoping to snag the seat of a no-show. Our spirits were crushed. As we turned to leave, Katelyn, who had been reluctant to "go see some guy just talking," said, "Oh, I wanted to see how he helped the kids."

Jeff had heard him interviewed on NPR the day before and said, "He's lecturing all over the Bay Area. Maybe we can catch him somewhere else." But without Internet access, we were unable to find out his speaking schedule. A couple days later, I called Kepler's Books, who had sponsored the lecture, and the clerk offered to look it up online, but by that time he was in Sacramento already.

Day 238: As Katelyn began her three-page letter to Santa, she insisted that she be able to ask for three (not two) presents from Santa, "Because I've been such a good girl, ya know." And added, "Plus, you and Daddy are going to get me presents too, right?"

"Of course, my love. What would you like?" I asked.

"Well, from you and Daddy, I'd like some of those things you put in your ears -- you know, so you can hear your iPod."

This request is coming from a girl who hasn't even seen her iPod in nearly seven months, plus she already owns some ear buds, since they conveniently come with the iPod.

"They're called ear buds," Jeff added. "So, what do you want from Santa? You need to get your letter done because it's almost bedtime."

The three of us sat at the kitchen table in silence as Katelyn wracked her brain for the perfect gift. The letter was like a magic lantern and Santa was the genie -- waiting to grant wishes to children worldwide.

Katelyn, who had not been at a loss of what to ask for Christmas for the past three months, was suddenly stumped. She had already forgotten about just how badly she needed the Barbie dollhouse, the magic set, all the things that a week ago were "must haves."

Finally, she blurted out, "I'm going to ask Santa for diamond jewelry -- real ones, not pretend. A ring, a necklace, and a bracelet -- all matching."

My eyes grew as big as oranges as I looked over at Jeff. Real diamonds were out of the question. Luckily Katelyn wouldn't know the difference. Finding a bracelet and ring to fit a child would be tough. Obviously, she had seen too many Macy's ads.

"How about just a necklace? We want to leave room in Santa's sleigh for presents for other children, too," I said. She agreed.

"How about a K-shaped necklace with diamonds?" I suggested, having seen one at Kohl's recently.

"Oh, yes!" she said.

I drew a sketch of a K with small round diamonds adorning it. "Like this?"

"Oh, no. I don't want diamonds *on* it, I want it to be one solid diamond in the shape of a K…and how about on the back it can have my birthdate -- 'born 2003,'" she said.

Here we go again. Last year she wanted a music box with a twirling ballerina. But it had to be pink, with a mirror and two drawers. She had me running all over town. God forbid Santa doesn't get her specifically what she asked for! Santa cannot let her down.

"Santa's very busy," said Jeff. "I don't know that he'll have time to make something that elaborate."

"Oh, sure he will. He's Santa," she insisted.

"Well, why don't you write a K diamond necklace -- and I'm sure he'll get you one that he thinks you'd like. "

After nearly an hour, Katelyn's letter to Santa was finished.

"Now what else would you like from Daddy and me?" I asked.

"The Toys "R" Us ad had a Dora doll -- the teenager Dora, not the toddler Dora -- I'm too old for that," she said.

Dora is no longer just for preschoolers, now elementary school kids can pal around with her as well, thanks to Nickelodeon's new marketing ploy. It turns out that customers could "customize" the new Dora online – changing her eye color, jewelry color, and hair length with the click of a mouse. Katelyn didn't have to even watch the new grown-up Dora show to be sucked into the merchandise.

"I don' know if toys are the best things," Jeff said. "You don't even have time to play with the ones you already have."

"Oh, Daddy. It's Christmas. Of course I need toys," she said, batting her baby blue eyes at him.

Looks like Dora would be spending Christmas with us.

Day 240: I whisked Katelyn off to Macy's to drop her Santa letter in the special mailbox and managed to catch a glimpse of the jolly old elf himself. A newspaper ad informed us that for each letter dropped in the mailbox, May's would donate one-dollar to charity. Katelyn's reaction was beyond her years.

"They're just trying to get our money," said Katelyn, pulling open the handle to the mailbox and letting her letter slide in.

"No, it's free," I insisted. "You don't pay to give them your letter."

"No, but they just want you in the store so you buy stuff," she said.

My jaw landed on my lap at the thought of Katelyn's insight. I hadn't even thought of that. Here I was just focusing on helping others, when in fact Macy's was probably making at least twenty dollars spent on every one dollar donated, as is evident by the fact that I purchased a wallet for Jeff before leaving the store. How could I resist, it was forty percent off!

Month Nine: It's a Wonderful (Wireless) Life

"Technological progress has merely provided us with more efficient means for going backwards."
--Aldous Huxley, Writer and philosopher

Day 241: At breakfast with my friend Macy, the subject of Santa came up.

"I'm afraid Katelyn's on the verge of not believing, but I'm never going to admit there is not Santa. It's a lie that I intend to keep telling," I said.

"Yeah, me too. My parents never gave us the chance to believe in Santa. They were always up front with us. I was the one in school telling all the other kids that there is no Santa. But then sometimes my aunt would put a gift from Santa under the tree and I knew it was something my parents would never buy, like a Barbie. And I did wonder if it really was from Santa," she said.

"Oh, that's terrible," I agreed. "Yeah, I hate how so many TV shows and movies have a character that doesn't believe. That just ruins it for kids because then they start questioning it. So why didn't you get Barbies?"

As I dig into my syrup-drenched pancakes, I listened intently to Macy's story.

"My parents were very anti-commercialism, so I wasn't allowed to have dolls that already had names, like Barbie or Winnie the Pooh. Instead, I was given ragdolls that I could name myself. I always thought, 'Can't you just give me the Barbie and I'll give her a new name.'"

"I can see where they're coming from," I told her. "I had no idea how susceptible our family was to advertising until we unplugged. I mean Katelyn has more that 20 Barbies, everything princess and *Hannah Montana*. It doesn't allow for as much creativity for kids. It's not like I wouldn't buy Katelyn a brand-name toy. Now, I'm just more conscious of it."

Day 242: As Katelyn put the finishing touches on the sugar cookies, dusting them with colored sprinkles, she licked frosting from her fingers, sat back and smiled at her masterpieces.

"The neighbors are going to love these," she proclaimed.

Our annual holiday cookie giving was expanded from just friends to neighbors. With less TV, we had more time to bake. With less TV, we had been given the opportunity to get to know our neighbors a bit better over the past few months. Wanting to further foster these relationships, we filled cellophane bags with sweet cookies and tied it with curling ribbon. Feeling a little Santa-ish myself, we gathered our gifts and doled them out to our unexpecting and appreciative neighbors.

Normally reserved, Katelyn bounced from house to house gladly spreading the holiday cheer. One woman, who lives next door, had recently lost her husband and her face lit up like a Christmas tree at the sight of Katelyn on her doorstep -- a little Christmas elf. She warmly invited us into her home for the first time, to show Katelyn her purple themed Christmas decorations and her tree strewn with brightly colored LED lights. She was a grandmotherly figure that Katelyn seemed comfortable with. Upon leaving, she invited us over for her holiday party the following weekend. We graciously accepted. Giving feels good.

Day 244: Our neighbor Hillary rang our doorbell. I was greeted with a platter of assorted homemade cookies. "I was making some for my (grown) kids and thought you'd enjoy some too." She bent down to Katelyn's eye level and said, "By the way, the ones you made for me were delicious."

A smile crept up on Katelyn's face. I closed the door and carried the delectables to the kitchen. Katelyn followed close behind, anxious to get her little fingers on some fudge. Indulging in a piece of chocolate heaven, she said, "It's so nice how we gave cookies to the neighbors and then got some back. It's nice to communicate with the neighbors."

"Yes, it is my love."

Day 245: Christmas with Grandma and Grandpa meant a pretend Barbie Blackberry for Katelyn.

"Mama, when I grown up, I want to get a real phone like this. By that time I'll be so good at texting," she says as her fingers tap away at the tiny pink buttons.

"I just looked on the Disneyland website and they are building a whole new section next year," she announces, using her imagination.

With icons that represent e-mail, the Internet, a calendar and an address book, it is sure to keep Katelyn entertained with hours of tech-related tasks.

"Oh good, now I can buy my American Girl online. Let me look up the website. Oh, they need my credit card number," she types in her fake number.

A few seconds later she switches to the phone mode and is placing her order with the operator.

"What happened? The website didn't work?" I ask with interest.

"Oh, no, it just showed what they had, but then you have to call to order."

"Make sure you give them your shipping address," I play along.

"They already have it," she nods, phone to ear.

"How?" I inquire.

"It's magic. They just know where to send stuff."

Day 248: As I walked through the front door, Jeff greeted me with a sad face.

"I just got an ear-full from my mom about us not being on e-mail," he informs me.

"What's going on?" I say with concern.

"Apparently my grandmother was very touched by our last conversation with her and she wanted to contact us, so she sent us an e-mail since she can't hear well on the phone. When she got our bounce-back she was confused so she asked my mom what it meant. When my mom explained to her that we were unplugged, she started to cry."

"Oh, no. I'm so sorry," I say. "It's not our intention to upset your 90-year-old grandmother."

"My mom is not happy about it and in her mind there really is no rational explanation for us not being accessible via e-mail."

"To tell you the truth, even when we do plug in, I don't think I'll return to e-mail," I admit.

"How can you say that? I can't wait to get back on e-mail. I miss being in touch with my family," Jeff replies.

"I think it's different for moms," I tell him. "Getting back on e-mail will be like opening a can of worms. If I tell everyone I'm back online, they'll start e-mailing me again and, let me tell you, moms like to e-mail. It's not uncommon to get a couple dozen e-mails daily. I just don't want to be tied down to e-mail anymore. And the thing is, if I check it and reply to one, and not another, and the word gets out that I'm online and didn't reply to someone's e-mail, then feelings would get hurt. Once you're online there's an obligation -- an unwritten rule that you have to reply to every e-mail you receive. It's not how I want to spend me time."

"Well, what if you just check your e-mail once a week?" he suggests.

"That wouldn't work. You know how e-mail is- --people expect an immediate reply. I hate that I feel pressured to conform to what others want me to do."

"So what are we going to do to smooth things over with my mom and grandma?"

"We will make an effort to send her letters and photos," I said.

"And I can draw her pictures to send her," Katelyn added to the conversation.

"Perfect. And we'll buy her some pretty stationery, a nice pen and stamps so she can write back," I added.

"I'm not sure if she can write. She's pretty old," Jeff said with concern.

"I have an idea. It could be a bonding moment for your mom and grandma. Your grandma could dictate and your mom could write for her. That could be a special time for them."

Upon informing Jeff's mom of our suggestions she seemed pleased. I argued that it would be more personal and meaningful than e-mail.

Day 249: Christmas Eve and Katelyn was tucked in dreaming of sugarplums. Jeff and I were knee deep in wrapping paper and bows. As excited as I was about seeing Katelyn's face on Christmas morning as she discovers all the gifts waiting for her under the tree, I was even more excited to give a gift to someone else, Macy. After hearing how she was deprived of the whole childhood Santa experience, I felt compelled to give her a taste of the magic of Christmas. While shopping, I had come across the perfect gift -- a silver ring inscribed with the word, "Believe." Early in the evening, my attempt to stash the gift under her tree failed. Masked by the deed of delivering cookies, I just couldn't get up the nerve to fling the ring box under the tree. So instead, dressed in my flannel PJs and robe, I hopped into the car after playing Santa for Katelyn and drove off into the dark of the night to deliver the last gift.

With my Prius as my sleigh, I drove silently through the deserted streets. Arriving at Macy's, I dashed to the porch, made the drop and gleefully hopped back in the car and headed home, hoping I had gone undetected. If only I could be there to see her look of surprise when she discovers the gift from Santa.

Day 251: This year Christmas took on a new meaning. We were subjected to the Toys "R" Us ads in the newspaper, but without the TV we were not bombarded with the heavy holiday advertising. We decided to focus on less materialistic, more sustainable gifts. For Jeff's parents, cloth napkins. For the cousins, bird feeders. For Katelyn's teacher and our parents, a flock of chicks through Heifer International. For my parents, tree ornaments made by Katelyn. Wind chimes for me -- to drown out barking dog; chin-up bar for Jeff; keyboard for Katelyn to learn music on -- Dora, Barbie and Cinderella didn't join us this year.

After reading about the amount of wrapping paper in landfills, I had planned on using comics to wrap gifts, but like a siren's song, I got hypnotized by the glittery paper and bows being sold at every store and couldn't resist. Starting out with good intentions, I did wrap a couple gifts with comics. One day Katelyn walked past the recycling bin in the kitchen and noticed the Sunday comics mixed in with the rest of the paper. She scolded me, "Mama, this isn't trash. It's wrapping paper."

Without my digital camera to take the perfect family portrait, we decided to go low technology and do something different for our Christmas cards. I had Katelyn make a drawing that we had made into cards.

The biggest disappointment this year was that I wasn't able to make our traditional photo calendar for Jeff. My solution -- I found a calendar that allowed Katelyn to draw pictures for each month. A touching alternative.

Another missing piece was that it was the only Christmas not captured on video. A pivotal Christmas -- Katelyn started to question whether there was a Santa or not. Those years of believing are so few. *Would this be our last magical Christmas?*

Typically, our "big" gifts would have been the latest high-tech gadgets. Maybe next year we could be really green and see if we could buy all of our gift used, off Craigslist or eBay, to save the millions of tons of trash from going to the landfill.

Day 252: Checking in with one of my nature-loving friends, Heather, I was shocked when she revealed, "Santa brought our family a Wii." Even Heather, who always kept a tight rein on screen time, gave into the allure of video games.

"First we just allowed the kids to play Wii on the weekends, but then that's all they wanted to do, and it sparked an increased interest in them wanting to play more computer games," she admitted. "We've always allowed one hour of computer time a day, but it got to the point that they were rushing through their homework just to get to their games. So we had to limit computer time to weekends only, too."

"I think it's easier just not to have it at all," I said.

"Yeah, we told the kids that we didn't approve of Santa's gift, but if they could agree to use it only occasionally then we'd give it a try," she said.

Wii has definitely been an improvement in the realm of video games because at least it addresses the sedentary video game playing. But it still feeds the addiction that is robbing kids (and adults) of the outdoors. What's next, a Wii jogging game? Who needs a treadmill or a trail, we'll just jog on a virtual trail without the bother of sunscreen, bugs, crowded paths -- let's just never leave the house.

Why doesn't Wii use their power for a good cause – create a game that teaches and promotes environmental education while being active and ends the game with encouraging users to get outside. Maybe a game where each point earned, Nintendo will donate to a good cause

"I can't believe you're still unplugged," Jaime said. "You are the only person I know without e-mail."

"It's very liberating. You should try it," I tell her.

"I would love to but I need to be online for school, Mom's club and work," she says.

"You'd be surprised...once people realize you're not online, if they need to reach you, they'll find a way. As far as I know, I really haven't missed much by not being on e-mail. I am proof that it is possible to live a full, rich life without e-mail," I joke.

Day 253: Christmas has come and gone and we managed to make the season merry even without a TV. Rather than sitting idly immersed in watching Rudolph, Frosty or Santa relive the same story as last year, I made a conscious effort to partake in local holiday festivities with the family.

Thanks to the Saratoga library, we were treated to a puppet show (The Night Before Chris Mouse), a marionette show (Winter Follies) -- featuring an ice skating snowman, and a magic show starring a world-renown magician. Turns out he was the ringmaster for the recent Barnum and Bailey circus we had taken Katelyn to. Mr. Zing Zang himself was giving back to the community he grew up in. "It's much nicer to see the reactions of the children's faces up close, rather than with six spot lights in my eyes," he commend while entertaining us with his slights of hand.

After the show I asked Katelyn, "Wasn't that better than staying home and watching TV?"

"Yeah, but I still miss it," she said

The "piece de resistance" for our Christmas outings was a trip to Bethlehem -- complete with live camels, cows, sheep and, of course, a screaming baby Jesus wrapped in swaddling fleece. Sponsored by a local church, a vacant parking lot had been transformed into the birthplace of the most famous man in history. Visitors waited in a line more than 10 blocks long to make the pilgrimage while King Herod rode through the streets in his chariot spewing insults to those wishing to pay their respects to the baby. Not having the patience to withstand the wait, we walked to front of the line to peer in through the entrance. As we tried to get a glimpse of the bustling Bethlehem, a Roman guard took pity on us and waived us through. People were paying their taxes, playing dreidel, stopping at the farmer's market, making pottery, cooking over the fire, dancing in the streets.

In a corner stood a manger covered in fronds while Angels hovered above exalting their King. The winged beauties sang in unison from the rooftop. As Mary and Joseph cradled baby Jesus beneath them, men, women and children crowded around to catch a peek at the precious babe. This was the true meaning of Christmas. Simpler times. Love. Humanity. Saving grace. I could only hope that years from now there would be an exhibit of Christmas past featuring mountains of presents, piles of packaging and wrapping paper, and mounds of overindulgent processed foods. Visitors would whisper, "To think Christmas used to be about gifts not family. And look at all the trash it produced. And how sad that they didn't eat food fresh from their garden."

Maybe someday we'd get Christmas right.

Day 254: Christmas was over, New Year's Eve was yet to be. There was a lull in our house. Jeff was off work, Katelyn was out of school and we were bored.

"Let's go skiing," Jeff said. "See if the Tahoe cabin is available."

Granting his wish, I made the call. No luck. It was booked.

"How about Yosemite?" I said. "They have a small ski resort and we've never taken Katelyn to the Valley. It would be so beautiful. Let's go for New Year's Eve!"

"I just really wanted to go to Tahoe," Jeff pouted, wanting to ski the steep runs that Squaw Valley has to offer.

"Might be fun to go somewhere different," I said. "I'll call and just see what accommodations are available."

The choices were three hundred dollars for a room at the four-diamond Awahnee Hotel or eighty-six dollars for a heated canvas tent cabin.

"It'll be like camping," I said, knowing that was something our family always enjoyed.

"Sounds like a lot of work," Jeff shook his head. "I didn't want to have to walk to the bathroom in the cold of the night."

"What? This is coming from Mr. 'I go backpacking for days in the Sierra.'"

"But it's New Year's Eve," whined Jeff. "What are we going to do after dinner? Just go back to the cabin and sleep? I was hoping we could put Katelyn to bed, share a glass of wine and watch the ball drop on TV. There are no TVs in the canvas cabin."

I chuckled since Jeff hasn't made it past the strike of ten on New Year's Eve since becoming a parent. I have spent the past few years ringing in the New Year with Ryan Seacrest. This year would be different. Pre-Katelyn our New Year's Eves were spent sipping fine champagne at swanky San Francisco bars surrounded by friends, decked out in our favorite duds.

Upon Katelyn's birth, we came to the grave realization that finding a babysitter for New Year's Eve would be harder than winning the lottery.

After letting Jeff's pouty-puss mood ruin my day, I decided to make an executive decision. I booked the tent. He was going to go to one of the most beautiful places on earth and actually enjoy, damn it.

After finalizing our itinerary I made the announcement. "We're going to Yosemite for New Year's Eve. It will be part luxury (cocktails at the Awahnee) and part rustic (sleeping in a tent)."

Having recently learned of our family connection to the Sierra Club, Yosemite seemed an appropriate place to celebrate the start of a New Year. If we did have our TV, I assume we would have settled for the norm. Once again without it, we're forced to create new experiences.

Day 255: During the four-hour drive to Yosemite, Jeff was chauffeur while Katelyn and I sat quietly in the back seat. This was the first low-anxiety road trip since Katelyn was born. She sat comfortably next to me reading a *Magic Tree House* book -- gifts from Santa -- while I indulged in my latest choice of literature.

No whining, no fussing, no fidgeting, no movies, no entertaining. Just peace and beautiful countryside. As we edged closer to the park, snow dusted the ground like powdered sugar. We'd have a white New Year's Eve. Braving the 32-degree weather, our first stop, once inside the gates, was a stroll up to Bridleveil Falls (our toes tucked snuggly into snow boots). Katelyn stood in awe, peering up at the rush of water looming overhead.

"Wow, Mama. That's a huge waterfall," she said, her breath warming the air. Only a handful of tourists gazed at the site. Winter weather kept crowds away. During the summer months, Yosemite welcomed swarms of visitors, bears and bees.

Continuing down the scenic road, we arrived at Curry Village and checked into our modest accommodation -- a heated canvas tent cabin. Wheeling our luggage through the hard-packed snow, I couldn't help but wonder if Jeff was longing for a hotel room with valet parking and a bellboy -- something he'd grown accustomed to on his business trips.

I inserted the key into the padlock that secured our temporary home. *Click*, we were in. The warmth of the cabin poured out, penetrating the crisp air that entrenched us. I flicked on the light switch to illuminate our abode -- double bed, two singles and a small dresser. A step up from camping.

Upon check-in, we couldn't help but notice the plethora of signs warning us of bears -- as much of a tourist attraction as the waterfalls and granite walls. We had to sign a waiver agreeing to store food in the bear box, not in the tent. And for added emphasis, a video of bears peeling apart cars to get at food appeared on a screen behind the front desk as the clerk cheerily handed us our keys.

After meandering through the visitor's center, we were driving back to Curry Village for a slice of greasy pizza by a roaring fire. As we rounded a corner, what to our wondering eyes should appear -- a full moon rising majestically over Half Dome -- the most recognizable

Yosemite landmark. We pulled the car to the side of the road and watched, mesmerized by the natural wonder. Not only was it a full moon, but a blue moon.

I grabbed my 35mm camera and perched it on top of the car since my tripod was packed away in the bear box. Soon, the roadside was dotted with red break lights and tripods were out in full force. Digital cameras clicked around me, followed by "oohs" and "awes" as photographers stared at their LCDs.

While packing for our trip Katelyn said, "Make sure we bring my camera so I can take pictures of our trip like last time." She was referring to her *Hannah Montana* digital camera that we purchased before our springtime trip to the Grand Canyon -- as a way to entertain herself during our hikes.

Since the camera didn't jive with our unplugged lifestyle, I quickly countered, "Oh, I bought you a disposable film camera. It's all yours. You can take photos of whatever you like (up to 36)." Before we arrived she snapped a photo of her small Christmas tree, daddy driving the car, Half Dome from the car. I didn't have the heart to tell her that her images would not turn out if taken from a moving car, at night without flash, of a moving object, or pointing at the sun.

After retrieving my tripod, we set out for a moonlight hike and ice-skating under the stars. As we walked down the path, flashlights in hand, we gazed out at the snowy meadow surround by granite peaks. A layer of fog blanketed the valley floor as if God was tucking the woodland creatures in for the night, turning on the full moon as a nightlight. The scenery was surreal as we stood in darkness, watching our breath get swallowed by the fog. The night was silent.

The stars twinkled above as the moonlit snow glistened below. I expected the sugar plum fairy to come waltzing out from behind a Sequoia at any moment -- pirouetting across the frozen glade. As the temperature dipped below freezing, skating on ice was the last thing that appealed to me, but Katelyn -- who for the longest time had career aspirations of becoming an ice skating cheerleader -- was hard to dissuade.

Jeff, not afraid to admit his lack of ice skating skill, was the official photographer. As Katelyn and I twirled around the rink, hand-in-hand, Daddy captured the moment on film. This was Katelyn's first trip to Yosemite Valley -- something I'm sure she'd always remember. If only we would've been able to capture it with video. This age, this time, this place warranted a video camera. It was a moment in time we'd never get back and could never relive by watching the video.

Photos and videos have always been a high priority to me -- probably too much so. The week before we unplugged, we took a road trip across the Southwest (a first for us) -- stopping at Arches, Bryce,

Zion, Grand Canyon, Sedona and Scottsdale. Due to our dedication to capturing the moment, we spent parts of our vacation at Target (replacing our broken video camera), Radio Shack (buying batteries), Wal-Mart (charging our camera), Radio Shack again (buying flash cards), arranging to have FedEx send a left behind charger from one hotel to the next, uploading hundreds of photos, posting nightly blogs about our trip, and writing articles for online publications every night. We were no longer slaves to technology, but rather, we were forced to live in the moment and soak up our surroundings and relish our time together.

Day 256: Jeff spent the better part of the night scampering through the snow back and forth to the bathroom, struck by a bought of stomach virus. I couldn't help but see the irony in the situation. His biggest complaint "not having a bathroom in our room" became his biggest nightmare as he trudged the 50 steps to and from in the bitter night.

The next day was spent perusing the Yosemite Museum, Indian Village and Ansel Adams Gallery. Katelyn happily purchased an Indian-inspired necklace. Jeff and I were eager to see the Ansel gallery since his great grandfather was a friend of his. We were disappointed to see it was very commercialized with postcards and posters, rather than more original art.

Before dinner, in preparation for the New Year's Eve festivities, we settled down for a short winters nap. Next up, a leisurely walk to one of the world's tallest waterfalls -- at 2,425 feet -- Yosemite Falls. While trenching through the snow, Katelyn squeezed my hand and said, "Mama, I still do miss TV a little, but I like how we've been doing so many things outdoors lately and learning about the old days."

As I looked down at her with her golden locks flowing from beneath a fleece cap, I breathed a long sigh of relief realizing the past few months of struggling without technology was worth it. Glancing over at Jeff, he smiled warmly as I bent down and hugged Katelyn tightly and whispered in her ear, "Thank you."

We emerged from our tent dressed in our New Year's Eve best -- with snow boots and down jackets. Splurging on the two hundred dollar tickets to the black-tie soiree at the Awahnee was out of the question, so we opted to voyeurs instead. Katelyn had never seen so many women in fancy gowns, made from sequins and silk. The tuxedo-laden men escorted their wives. The expansive lobby was warm with a subdued sense of excitement. The stone fireplace, nearly big enough to hold a Mini Cooper, cast an orange glow on the partygoers. Red wine and red gowns flowed elegantly and effortlessly.

We left the elegance of the Awahnee and were thrust into reality of parenting, at the kid's party at the Yosemite Lodge. A small room, lit with fluorescent lights, jam-packed with children screaming with excitement while others were mid-tantrum from no nap, too much sugar and being up past bedtime. The floor was strewn with confetti, tables covered with plastic 2010 tiaras, glow in the dark jewelry and a buffet that consisted of pretzels, cookies, chips, and sparkling cider. The overstimulation factor was off the charts. Katelyn watched in horror at the chaotic scene and begged to be removed. We happily obliged.

While impatiently waiting for our nine-thirty dinner reservation, we grabbed the only seat we could find in the bar -- on the edge of the fireplace. A glass of wine for me, Shirley Temple for Katelyn and water for Jeff and his upset tummy. At a few minutes before nine, my eye caught the two large screen TVs hanging precariously over the bar.

As the bartender busily filled glasses with libations I inquired, "Is there somewhere to watch the ball drop at midnight?"

"Oh, great idea," he said searching for the remote. With a flick of the button, the TVs came to life. Thanks to the magic of live satellite TV, we got to celebrate midnight in New York City. Everyone in the bar counted down in unison - *Happy New Year. Does that mean we can go to bed now?* I gave Jeff a peck on the cheek, "You got to watch the ball drop after all!" This was the most festive New Year's Eve in years.

Once seated in the restaurant, Jeff and I dined on overpriced fare, as Katelyn napped, her head on my lap and legs, sprawled along the booth seat. We lingered over our meal, enjoying adult conversation. The dishes were cleared, the bill paid, and we headed off to our cabin in the woods -- a sleepyhead in tow.

After brushing teeth and donning jammies -- it was nearly midnight and Katelyn caught her second wind -- determined to ring in the New Year. She dispensed the New Year's Eve tiaras, tinsel leis, and party horns. At the stroke of midnight, we shot some confetti poppers in our cabin. Wished each other a happy 2010 and lights out.

Day 257: The first day of 2010 was spent on the slopes -- Katelyn's second season on skis and she was tearing up the Bunny Hill at Badger Pass. The sky was filled with falling snow. Katelyn held out her tongue as a landing pad for stray snowflakes that dissolved instantly on impact. As we frolicked in the winter wonderland, we were thankful for our life unplugged, but concerned that 2010 would also be the year we would plug back in. *Or would we?*

Day 258: As the morning light streamed in through my bedroom window, I blinked my eyes open in protest -- not eager to start my day at such an early hour. I'm convinced Katelyn is part rooster -- up before the sun each day, full of zeal, ready to tackle the day ahead of her with gusto.

"Mama," she taps my shoulder with her boney finger.

I roll over and see her, bright-eyed and standing over me -- looking like an angel with her long blond locks past her shoulders.

"What are we going to do today -- ice skating, rock climbing, hiking?"

It's Sunday morning and a New Year. Having spent an action-packed, few days in Yosemite, Katelyn is raring to go…then it strikes me. Last year, before we unplugged, our Sunday mornings began with Katelyn snuggled in our bed watching cartoons while Jeff and I dozed to a decent hour. This would be followed by Jeff and I trying to convince Katelyn to get out of the house with us. She would vehemently protest -- preferring the familiarity of her surroundings at home.

Katelyn is no longer an indoor child as she once was. She has grown into a confident adventure-seeking, nature lover in a matter of months. We no longer have to bribe or threaten her to get her to agree to go on a hike. She truly has undergone a metamorphosis. Now that she has spread her wings, I can't wait to see where the wind takes her.

Day 260: Julie arrived today, with her hairbrush. Katelyn adored her bellbottom jeans and embroidered tunic top. Julie is an American girl…doll, that is. The reward for Katelyn's hard work and patience. After having saved her allowance for nearly six months, her prized possession was finally in her hands. For over a year, when an American Girl catalog landed in our mailbox by mistake and I innocently gave it to Katelyn, she's been pining for a doll ever since. There was no way I was about to dish out over one hundred dollars for a doll that would inevitably sit alone in a corner of Katelyn's room after a couple of fun filled days, a la *Toy Story Two*. So I put Katelyn to work -- set the table, put away laundry, clean her room. After months of hard work, she managed to save ninety dollar. For Christmas, we gave her the additional funds needed to bring Julie home.

When the original catalog arrived in our mailbox, Katelyn and I flipped through it curiously. I was instantly drawn to Molly McIntire, from 1944. Her family supported the World War II efforts by growing Victory Gardens. *Victory Gardens?* Having been born in 1970 myself, I had never heard of a Victory Garden. I was intrigued, but without access to the Internet, I decided to file the term away in my memory bank until I could figure out a way to find out its meaning.

In the months to follow, without making a conscious effort, Victory Garden trivia seemed to be popping up everywhere…at Hidden Villa, in *Sunset* magazine, from Jeff's grandma. It was like when you want to buy a new car and once you decided which make and model you want, you suddenly see them everywhere on the road. As if we have an internal search engine in our minds -- once you focus on something, it becomes more apparent. That seems to be happening frequently since we've unplugged. But you need to slow down to notice it.

After learning that Victory Gardens were a way for people to save money during the great depression, it strengthened my resolve to promote gardening as a way to cut expenses during the current economic decline. Spend less, eat healthier, and build communities. As I looked out at my weathered winter garden, I was thankful to have been able to share it with my friends. If only I could build a website to post updates of the produce's progress.

Now that Julie was a part of our family, I asked Katelyn, "So, what are you going to buy next with your allowance?"

"A Nintendo DS," she said relentlessly.

Day 265: Jeff said something I thought I would never hear cross his lips, "Can I make dinner sometime this week?"

I shot him a glance with raised eyebrows in disbelief.

"You are going to cook…dinner?" I asked suspiciously.

"Sure, why not?" he said.

"Uh, because we've been married nearly ten years and I can count how many times you've offered to cook dinner on one hand."

He shrugged and said, "I thought maybe I could make fried rice."

"Ooh, yes Daddy. Like at Panda Express," Katelyn said licking her lips.

"Yeah, do we have a recipe?" he said opening cupboard doors randomly in search of a cookbook.

"Why do you need a recipe? It's just rice, eggs, peas and carrots," I said.

"If I'm going to do it, I want to do it right. I need a recipe," he insisted.

So, his offer to make dinner meant I had to be the one to take the time to find the recipe.

After flipping through a dozen cookbooks, I said, "Here's one for rice pilaf."

"That's not fried rice," he said.

"Close enough. Just throw in some eggs. It'll have to do. I don't exactly have access to the Food Network website to do a recipe search," I reminded him.

Ah, the Food Network. How I missed my personal chef, Rachael Ray and her handy 30-minute meals, a Godsend to overly schedule moms. Isn't it interesting how so many of us moms spend a half-hour watching a cooking show, then a half-hour to cook, when we really could've had that whole hour to prepare a more elaborate meal, if we wanted to. And nearly every mom I know uses the TV to keep the kids entertained while mom is busy whipping up a quick meal in the kitchen.

Now it depends on the size of the family and the temperament of the kids (and the mom's level of patience), but rather than shoving the kids away -- why not invite them into the kitchen to help prepare the meal -- scrubbing potatoes, peeling carrots, measuring ingredients. Create an evening ritual to let your children bond with you, instead of the TV.

In her book, *Animal, Vegetable, Miracle*, Barbara Kingsolver and her family spend a year eating only homegrown and local food, which leads to lots of quality family time in the garden and kitchen…not in front of a TV or computer eating processed foods. An experiment we could all grow from.

Month Ten: Seeking Silence

"It has become appallingly obvious that our technology has exceeded our humanity."
--Albert Einstein, Theoretical physicist

Day 272: I asked Sage if she'd like to borrow a great book I'd just finished. Her reply, typical of a mother of two under age seven, "I don't have time to read." Time -- that is one thing technology sucks from us, without us being fully aware. I would typically receive about 500 e-mails a month -- each needing a reply. That's 1,000 e-mails sent and received each month with an average of one minute spent on each, that equals 200 hours a year, at a minimum.

As for TV, I spent a minimum of three hours a day watching *Oprah,* the *Today Show*, *Entertainment Tonight*, *Rachael Ray*. That's 1,095 hours a year.

Don't even get me started on digital photography. With a minimum of 200 photos per event, uploading, editing enhancing, e-mailing, printing, posting. That's easily 144 hours a year.

Internet usage probably averaged about three hours a day using maps, looking up phone numbers, blogging, Google searches, parenting websites, Facebook, Craigslist -- just to name a few. That totals at least 1,095 hours a year.

With an iPhone, I would spend two hours a day making phone calls, playing games, checking my calendar, watching YouTube, utilizing useless apps. That would total 730 hours a year. With Apple reaching 1.5 Billion downloads on its App Store, I knew I wasn't the only one wasting time.

By my rough calculations, technology has robbed me of four-and-a-half months of my life per year.

Day 274: Katelyn plopped her backpack down on the kitchen table, unzipped it, and retrieved her homework folder.

"Here, Mama, this is for you," she said, handing me a piece of paper.

Scanning the note, I said, "Oh, it looks like the school is getting new computers for the media center. Have you ever used he computers at school?"

"No."

"I wonder if they are only for the older grades?" I said.

Curious to find out, I wrote Katelyn's teacher a note and slid it into Katelyn's homework folder.

Mrs. Nelson,

I noticed the school is getting new computers. Do the first graders get to use them?"

Kindly,

Sharael Kolberg

Her repliy arrived the next day via Katelyn's homework folder:

Dear Mrs. Kolberg,

Our assigned time to use the computers is when we also do our reading groups. At this time I feel reading takes precedence. Are you worried about Katelyn's computer skills?

Sincerely,

Mrs. Nelson

My return note:

No, she's been using a computer since she was two and has one of her own, although we're taking a break from it at the moment. I was just wondering who uses the media center computers -- if it's for certain grades.

She wrote back:

We can use them, but in the past they've been so outdated and would constantly crash -- it wasn't worth our time.

Parenting magazine quoted David Markus, editorial director of the George Lucas Educational Foundation, as saying, "...research shows that technology is highly effective in helping children retain information. Plus, using technology to do research and present ideas is a skill they're (students) going to need to be competitive in the workplace."

I guess it's a good thing the school is getting new computers after all.

Day 275: My head is so full of snot that I feel like I'm underwater. I can't breathe out of my left nostril. I'm congested and cranky. Katelyn is curled up with me, reading a *Magic Tree House* book -- enthralled with the character's mission. *How will they escape the current peril?*

As I moan in misery, I am so grateful for Katelyn's growing interest in books. She's been lying with me for over three hours. I snuggle up to her and ask, "What do you like better, books or TV?"

"TV because it has pictures," she says, without taking her eyes off her book.

"Yeah, but with books you get to use your imagination and create the pictures in your mind," I say.

"But there are things in books that I don't know what they look like -- like a coral reef," Katelyn replies.

Good point. You have to have seen it at least once to be able to imagine it.

"And I don't really know what the magic tree house looks like," she says with a bit of frustration.

Day 281: "Silence is golden" -- so the saying goes. To me, silence is irritating, uncomfortable and boring. Before we unplugged, and for the past 39 years, my life has been filled with a constant buzz from TV, radio, CDs, iTunes, iPods. To sit home alone in silence was just not possible. The constant need for stimulation was vital. It has taken me nine months to ease out of the uneasiness of silence. Without the luxury of a TV, I can no longer indulge in continuous "background noise." *Why in the world would I have wanted to fill my life with noise?* My venture into a less noisy life continued when a few months ago Katelyn started requesting that I turn off the radio while riding in the car.

"Can we just have some quiet time," she would say.

I bargained with her, "OK, but I get to listen to the radio at home then, just not while we're in the car."

Not rocking out to the radio while cruising down Hwy 85 made running our errands a little less jovial. I mean, come on, that is what we do. Hollywood can attest to it -- it's almost rare to see a romantic comedy without a car-singing scene. But no, we now drive without Lady Gaga or Taylor Swift to entertain us.

With the music off, Katelyn, much to my amusement, has become quite the chatterbox. Without my prompting, after school, she gladly recounts the play-by-play politics of the playground. I have learned so much about the social nuances of first graders on our five-minute drive home. Silence flung open a door to communication from the backseat.

No longer is it, "How was your day?"

"Fine."

Now, she rattles off the most minute details about her day -- Sandy cut her finger, Dylan has a baby sister, I went potty at lunch, Wanda was bragging about her new shoes, Ryan wanted me to chase him at recess, I picked out a new book at the school library. I had no idea how much of Katelyn's life I had been missing out on. And all it took was the flick of a switch to turn the radio off.

***

Katelyn raced off to her bedroom and was rummaging through her toy box. She emerged victoriously with her "laptop" -- a toy we had purchased for her when she was three. She powered it up, connected the mouse and squinted at the three-by-six inch black and white screen as it welcomed her, "Hello Katelyn. Choose an activity."

I hesitated to allow a mock computer in our unplugged household, but resisted banning it since she's gone without for so many months -- it wouldn't hurt to throw her a bone.

Day 287: Hunkered down in my office, just my typewriter, and me I am interrupted by the phone ringing.

"Hello?"

"Hi, this is John Smith. I'm trying to locate a Neil Fisk…who was in the Air Force."

"Neil Feist, you mean?"

"Yeah, Feist."

"Uh, that's my dad, but he doesn't live at this number. Who is this again?"

"John Smith. I'm with the Air Force. We're putting together a reunion and I wanted to invited Neil Fisk."

"How did you get this number?"

"On the internet"

"What website?"

"WhitePages.com, just typed in Neil Feist and got this number"

"That's strange. My dad has never lived at this number. Do you have a pen? I can give you his number."

After reciting the number, the man replied, "Oh, yeah. I already called that number, but he wasn't home. That was listed on the whitepages.com, too."

Thanks to technology, there is no such thing as privacy anymore.

Day 289: The children screamed gleefully as the red wiggler worms squirmed across their dirt-covered hands. Tenderly plucking the worms from piles of compost, the children were selecting which worms would be taking up residence in our new worm-composting bin at Katelyn's school. Prior to our "worm sort," I had arranged for the Santa Clara County Home Composting Education Division to give a worm composting assembly to the entire first grade. The children resembled a bin of worms as they squirmed restlessly while listening to the lecture on the cold cafeteria floor. It was an important lesson -- how worm poop is good for the garden.

Since my foray into gardening, composting was something that piqued my interest, but always seemed intimidating. I'm not sure if it was my lack of enthusiasm for dealing with the slimy creatures, or the rotting food to feed them. Eventually a series of events brought me to this day, bonding with worms in the school garden, with joy.

If I had not unplugged, I wouldn't have ended up volunteering at Hidden Villa. If I had not volunteered at Hidden Villa, I wouldn't have taken an interest in gardening. If I hadn't been interested in gardening, I wouldn't have visited Green Gulch Farm and Zen Center. If I hadn't visited Green Gulch, I wouldn't have wanted to learn more about meditation. If I hadn't wanted to find out more about meditation, I wouldn't have noticed or taken a flyer for a local "family meditation" session. If I hadn't taken the flyer and attended the meditation, I never would have seen the worm-composting brochure from the County of Santa Clara. If I hadn't called the number on the brochure, the Carlson kids might never have learned the value and importance of worm composting. It all started with pulling the plug.

When the lead school gardener had to step down due to working 12-hour days at her desk job, I gladly stepped into the roll of school garden CEO. I knew there was more opportunity than just having the kids digging in the dirt. But without Internet or e-mail, where would I begin…with my mother-in-law. Her friend ran the garden at Top of the World Elementary School in Laguna Beach. She's Jeff's former second grade teacher. When I contacted her, she mailed me a slew of information about school gardens.

From then, things just naturally fell into place. Hidden Villa provided me with additional lesson plans and I was able to write a curriculum (on my typewriter) for the remainder of the year at Katelyn's school. Next -- funding. *How would that come together?*

Day 292: As Jeff and I approached our 10-year wedding anniversary in May, my wish was for us to renew our wedding vows so that Katelyn (who was born after we got married and who had never even seen our wedding video) could relive our wedding day with us and take on the all-important role of flower girl. Our original plan was to fly off to Tahiti, where we spent our honeymoon. But given the economy and our renewed emphasis on relationships, it seems more important to do something local with family and friends instead.

After months of brainstorming, phone calls, and budget planning, we decided on the perfect place…Hidden Villa. We would rent the Dana Center overlooking a meadow and invite 50 guests. Most of our close Bay Area friends we met after having Katelyn, therefore they had not been at our wedding either. It would be nice to share this special moment with those who meant the most to us.

At the rental office, I met with Dyni to enquire about pricing.

"Do you have a brochure?"

"No, but I can e-mail one to you," she offered.

"Oh, I'm not on e-mail."

Digging through her filing cabinet, she commented, "Hmm. We might have a copy in here…somewhere."

No luck. I had to write down the information with pen and paper.

"So, I'll e-mail you the contract and you can drop it off with the deposit," she said.

"Uh, can I just pick up a copy now?" (*Since I just told you that I don't have e-mail!)*

"Oh, right. I can print one out for you from my e-mail."

"What about food? Can we bring our own or do we need to use your caterer?"

"We have a list of caterers, but it's up to you. You don't have to use them if you don't want to."

"Can I get the list?"

"Sure, I'll e-mail it to you."

"Uh, yeah. I'm not big on e-mail."

She stops, looks at me, Katelyn perched on my lap and smirks. In her European accent she says, "You're one of *those*. How do you say, 'odd bird.'" She chuckles and rifles through her filing cabinet again.

***

With our new techless lifestyle, Katelyn has become the "unplugged police."

"Daddy, are you supposed to be using your cell phone?"

"Mommy, shouldn't you be listening to a record instead of a CD?"

"Is it OK that we're watching TV while were at the restaurant?"

Our little rule follower.

***

Once we decided to unplug, I made the decision to put my journalism work on hold. I was lucky to have that luxury. And I realized that (when it comes to a career) for most of us, we truly can't unplug. Work must be done. Bills must be paid. If you are one of the millions that spend every day in a gray-walled, fluorescent-lit, artificially-aired cubicle, you can't cut loose from the ball-and-chain of technology while on the job. So, where does that leave those who desire to make the transition to a mini-tech lifestyle? Most cannot afford to make a drastic career change to something like waitressing, farming, retail, but it's worth mulling over. For those stuck in a career that requires constantly being plugged in, at least commit to rewarding yourself with a step outside the office a few times a day to give your carpel tunnel prone wrists a rest, breath some fresh air, and remember there is life beyond technology.

Day 299: Back to Double D's. This time it was a family affair -- the Olympics! The restaurant and bar were nearly standing room only and as loud as a rock concert. Arriving at 7 p.m., the supposed start of the opening ceremonies, the crowd was festive and lively -- definitely in the spirit. I felt as though we all should've been waving mini American flags as we wolfed down our Budweiser and French fries. Seated in a booth with views of TV screens surrounding us, we could not hear Bob Costas over the roar of the crowd, even though the TV volume was on full blast. The first clip we saw was of a bobsledder being flung off the track. Bob would say a few words, then the footage would replay, covering Katelyn's eyes when they cut to an image of the bobsledder's bloody face.

When the waitress came to refill our beer, I asked, "What happened to the bobsledder?"

"He didn't make it," she said, over the roar of the crowd.

So, they were airing video of someone being killed during the Olympics, and we did not have the luxury to fast forward the images. After that harrowing intro, I began to wonder who and why they would repeat such footage. Blood sells. As we lingered over our five-course meal of appetizers, hoping to keep our seats long enough to get a glimpse of the opening ceremonies, by eight-thirty, Katelyn was like a limp noodle and could not keep her eyes open a moment longer.

It was a significant difference from the 2000 Sydney Olympics where I watched the opening ceremonies from a live NBC feed aired on about 12 TV screens that lined a wall in a small dark room of the International Broadcast Center at the Olympic Village. As a media engineer for NBC, my colleagues and I digitally captured images from the live video feed and instantly posted them to the website. On my mandatory 10-min break, I walked out of the building and could see the Olympic Stadium as I watched fireworks explode overhead and heard the thousands of spectators and athletes cheer with anticipation of a memorable Olympic games. The energy was unforgettable. I spent the next eight weeks watching every sport from every angle for 12 hours a day -- glued to the monitors -- waiting for the picture-perfect shot. With an insider's perspective, I was privy to the fact that such an insignificant amount of actual coverage of the events would be aired. Even the gory stuff didn't make the cut: the high jumper with a pole in the crotch, the field hockey player with a stick in the eye, the weightlifter with the broken elbow, then foot. The absolute dedication and determination of the athletes, especially those whose sports were not mainstreamed, was astonishing.

After leaving Double D's, I yearned to witness whatever coverage NBC deemed worthy. That would be better than none. And at least I wouldn't be subjected to the live Facebook, Twitter and web updates to spoil the results for me. I needed an Olympic-viewing plan.

Month Eleven: From Computer Worm to Earthworm

"Our technological powers increase, but the side effects and potential hazards also escalate."
-- Alvin Toffler, American writer and futurist

Day 301: There is a continual flow of books in and out of my life. As I glanced at my desk, there was one borrowed from a neighbor, one sent from my mother-in-law, one purchased from the library, one packaged to send to a friend. Books have strengthened my relationships. I find myself drawn to certain types of books -- memoirs of hardships, environmental endeavors, non-fiction. Interestingly enough, I've noticed the people I am bonding with lately also find interest in this type of literature, therefore our conversations have become more deep and meaningful -- discussing global environmental policies, helping the underprivileged or finding meaning and gratitude in everyday events.

When I read *Three Cups of Tea*, I found it so inspiring and appealing to a wide audience. I started telling people about it as a "must read." Within a few weeks, it was read by Jeff (who never has time to read), my mother-in-law and father-in-law, my neighbor, and two close friends. Nice to have a topic of conversation with people from such varied backgrounds, all with their own views and opinions. Books make our lives richer. Thoughts and ideas are spreading person-to-person without the use of e-mail, Internet, Facebook or cell phones.

I can barely tolerate reading *People* magazine anymore. For one, it seems so shallow and unimportant (but, yes, entertaining) and for two, without a TV I don't know who half the people are anymore. The entertainment industry is so fickle that, in just a matter of months, I'm no longer up with pop culture. Doesn't seem to bother me though. *Do I really need to be up close and personal with someone named Snookie?* I'll stick with Michael Palin, Greg Mortenson and Richard Louv for now.

Day 302: As I dug my hand into the terra cotta pot and pushed the soft warm soil to the side, I slid the pansy in place. Happy in its new home, it was about to be joined by new neighbors, Mr. Asteria and Ms. Tulip. Jeff and Katelyn joined in the planting party as we transformed our front porch into a "welcome" garden -- a sight that would treat us, and our visitors, to a bountiful array of spring flowers each time we approached the entrance to our home. I couldn't think of a more romantic way to spend Valentine's Day. Jeff had forgone traditional cut roses and instead gifted me with pallets of petunias and various ornamental blossoms that would leave a much longer lasting impression. While my friends would be tossing their dried-up roses in the trash in a few days, my new garden would be a constant reminder of my loving husband for weeks to come.

In the past, I had presented Jeff with creative Valentine's Day gifts -- most of which were aided by the use of our access to technology -- my favorite being a customized CD I had ordered online in 1998 -- a compilation of love songs printed with a customized cover. This year Jeff and I were in synch. He smiled as I handed him a blueberry bush and Katelyn supplemented it with a strawberry plant. Our love blossomed over fruit and flowers.

Day 303: Katelyn sprawled out on our microfiber couch, flipping through one of her many American Girl catalogues. Lying on her tummy, with her knees bent and her toes wiggling in the air, she spouted out, "Mama, I really want to get the new American Girl doll, Lanie. She looks just like me."

Each year, American Girl has a doll of the year. For 2010, Lanie was supposed to be a representation of the current culture. Lanie's accessories included a laptop and her hobbies include exploring nature. As I peered over Katelyn's shoulder, leaning over the couch I said, "So, she comes with a computer, huh?"

"Yeah, isn't that cool? And she's just like me, she likes to be outside, like in her backyard."

"So, you can use a computer in nature?" I asked.

"Yeah, she's e-mailing her friends about how good worms are for the earth," Katelyn said seriously. So precious.

Day 305: The night was dark and damp. Raindrops drenched my windshield as teardrops soaked my face. Tears of joy and gratitude. As I maneuvered the windy road home, I was overwhelmed with a sense of purpose.

Tonight, Hidden Villa had hosted a volunteer appreciation dinner. Upon arriving at the farm in my mandatory spring colors, I entered the Dana Center and was greeted warmly by familiar and unfamiliar faces. People whom I had come to know over the past few months, that had become an extended family to me, gathered jovially to celebrate the spirit of volunteerism and nature. My fellow farmers had ditched their dirt-covered duds for floral patterns and pastels. I smiled seeing a more refined side of my down-to-earth friends. No matter how polished, there was still the telltale sign that this was a room full of farmers -- permanent dirt under fingernails. That was a right of passage I strived for.

For me, the evening marked a transition. I had spent the winter living happily as a caterpillar -- pulling weeds for David, befriending my fellow horticulturalists. Now, as we approached spring, I was about to emerge as a butterfly and take on a position that would let my true colors show. I was training with the Environmental Education Program at Hidden Villa and would soon be leading school kids on farm tours, teaching them about plants and animals and hopefully planting a seed in their minds about how they can protect our planet for future generations.

After an evening filled with meaningful conversations, farm-fresh food and green gifts, I drove home in silence, grateful that my lack of technology led me to become part of the farm family. I felt as though my inner self was finally being uncovered. After 15 years of being stifled by the constant distraction of TV, e-mail, cell phones, Internet and iPods, I felt the outer layer peeling away like an onion and my core finally free. I found me. It was a path that would bring positive change to me, my family, my community, and my earth.

Along with this self-realization came an overwhelming sadness for Jeff. Even though we were living a life without technology at home, he never has been able to experience a life unplugged due to the demands of his job. I longed for him to be able to escape the fast-paced, always online business world. I was eager yet hesitant to share my epiphany with him. There's a good chance that he would never be able to log off and slow down long enough to dig deep and unearth his true path in life. Technology has turned nine-to-five jobs into 24-hour jobs. There is no escape. There are no limits with the ability to be constantly accessible. Managers expect their employees to answer e-mails, phone calls and texts at any hour of any day. There is no such thing as vacation, holidays, weekends anymore.

Unless you utilize Timothy Ferriss's *4-Hour Work Week* strategy and outsource your work to virtual oversees assistants -- who will read and reply to your email for you, among other things.

***

"Hey Sharael! Thanks for coming!" Jake greeted me warmly, shaking my hand. "Should we sit outside? I always jump at the chance to be outdoors when I can."

Settling onto the cement wall surrounding the garden, I glanced down at my soil-stained shirt and khaki pants that converted easily into shorts with a zip. I tucked my dirty blonde hair under my baseball cap and adjusted my semi-scratched sunglasses. No resume, no polished suit, no conference room. This interview was far away from the corporate high-technology jobs I was used to.

"We're so glad you're interested in leading summer camp groups here at Hidden Villa," Jake said. "After discussing it with my manager, we'd like to offer you $275 a week with free childcare. Katelyn can attend camp all summer while you're working -- we'll waive the fee."

Did he say *per week*? *I'm used to making that much per hour!* Without access to Craigslist job listings, Yahoo jobs, Monster jobs, the pickings are slim. *Is $275 a week a competitive salary for this type of* *work? Can't do a salary search online.* I found myself saying, "I'll take it." Passion had trumped payment as I was about to transition from leading the schools group tours to being a summer camp leader.

***

As I reached into my medicine cabinet, my stomach dropped. I pushed the Advil and hairspray aside, to reveal my birth control pills. *Oh shit! What day is it? Friday!* I clutched my pills and realized I had somehow forgotten to start the new pack on the previous Sunday -- six days without birth control. For some 34-year-old women that have been married for nearly 10 years, that might not have been much of a concern. But for me, with our unplugged lifestyle that had made way for more sex, it caused panic. As much as I love kids, I was not embracing the thought of starting over with the sleepless nights, dirty diapers, and vomit-covered wardrobe.

I envisioned my life five years down the road without a kindergartener and middle schooler -- dealing with separation anxiety with kindergartener not wanting to separate from me, and me not wanting to separate from Katelyn. *Shit. Shit. Shit.* Too early for a pregnancy test, I would have to wait on pins and needles for three weeks!

Day 307: The sun bounced off the snow, like the reflection from a highly polished Porsche. Katelyn was comfortably perched atop her own skis, head in helmet, hands in gloves, feet in boots, eyes in goggles, ready to take on the mountain.

Unbeknownst to us, we had chosen to take our maiden voyage to Dodge Ridge on the Family Fun Festival weekend. Not only would we slush down the slopes, but we'd also be treated to viewing or participating in a variety of winter activities.

As the chair lift whisked Katelyn and Jeff to the top, they gazed down at the run below where school children flung themselves off jumps during the Big Air competition. On another run, parent/child teams raced down a giant slalom run, navigating through the gates to beat the clock. The bunny hill was converted into an obstacle course for the wee ones to navigate over, under, and around.

"Who needs to watch the Olympics on TV, we can see our own live Olympics right here," Katelyn announced, as she came to a snowy stop at the bottom of the hill.

Day 308: We finally brought home a pet today, well, actually 1,000 of them. We decided to forego the traditional dog, cat or guinea pig and instead opened our home up to garbage-eating worms. We are the proud owners of a "Wiggly Ranch," a.k.a. a worm-composting bin. Unlike traditional ranch real estate, we purchased ours from the city of San Jose for a mere fifty dollars. Then it was off to the Worm Dude to pick up our bag of gems.

I had been inquiring for months about where to get worms. Bins aren't so difficult to find with a trip to Home Depot, Smith and Hawken, and most nurseries. But worms -- not so easy to uncover. After asking everyone I could think of, I kept arriving at a dead end answer, "order them online." Easier said than done in my situation.

Then, one day the stars aligned and there, on the back of the worm composting brochure I grabbed at a meditation seminar, was the number for the Worm Dude -- with a 408 area code – my 'hood! Katelyn accompanied me to select our little wormies. After a 10-minute jaunt down Hwy 85, we arrived at the worm farmer's home. An Orowheat executive by day, Jerry (aka the Worm Dude) moonlighted as a vermiculturist. In his home garage, he was hosting nearly 100,000 worms.

For a burly man with large hands and a gruff voice, he had a tender nature and was very engaging with Katelyn -- making sure to meet her eyes when discussing ways in which she could help tend the worms. He dug his hand into one of several bins stacked on shelves head high and pulled out a handful of pale pink squiggly worms.

"Katelyn, look. Can you see the tiny baby ones?" I asked.

She nodded her head, a smile creeping from the corners of her mouth. He scooped some from each bin, showing us the variety of worms available. Like vampires, worms do not like light. They moved actively in his hand, looking like a live pile of spaghetti.

Next, he motioned us to see the finished product, lifting a pile of castings in his hand and said, "This is the worms' poop!"

He bent down to Katelyn and said, "Doesn't it look like dirt? Smell it."

To my surprise, she leaned over and put her nose ever so close to the poop and breathed in slowly.

"It doesn't' smell like poop does it?" he said to her.

Katelyn shook her head no.

"Put this on your garden and you're plants will grow big and strong," he said, brushing his hands clean and giving us specific instructions on how to care for our new friends.

"Feed them fruit, vegetables, newspaper, but not too much," he warned. "Keep them moist, but not too wet."

I felt like we should've had to sign a waiver or take an oath to "care for and protect" our wormies. Jerry, who got into the worm business after raising box turtles, seemed to have had a change of heart. He used to raise worms to feed to his turtles, now he was defending and protecting them. Those little squirmies wiggled their way into his heart. They have no eyes, nose, or teeth, but I swear they were smiling when Jerry lovingly pulled them out of bed.

"One of the best parts of raising worms is meeting the people that come to buy them. They are so 'down to earth' -- literally," Jerry said.

He handed over a white translucent bag, which I held up to see the shadows of our new pets. I left with a sense of responsibility and pride.

Upon arriving home, I handed Jeff the bag. He grabbed it innocently, "What's this?"

"Worms," I said grinning, "1,000 of them."

He quickly handed them back, squirmish at the thought.

"So, Katelyn have you thought of names for them all yet?" he joked. "Oh, maybe you can teach them tricks, like how to roll over."

Katelyn rolled her eyes with a grin.

Day 309: Time to put the worms to bed. Our newspaper strips had soaked overnight, turning into moist fluffy bedding.

"Cool, the worms can even eat their bed," I said as Katelyn looked on.

She giggled.

I filled he Wiggly Ranch tray with the newspaper, then came dinner -- eggshells, lettuce, banana peels, apple cores, coffee grounds. I handed the bag o' worms to Katelyn. She would take on the task of worm wrangler. Reaching out with her gloved hand, she gently loosened the bag, squealing with delight and apprehension as she peered inside at the mound of squirming worms. On the cold, dark backyard patio, I steadily held the flashlight as Katelyn cautiously placed the worms in their new home. She gently pinched the runaway worms between her thumb and forefinger and plopped them into place.

"It's too hard to grab them with my gloves on, plus it might hurt them. You have to be very gentle with them, Mama. They're fragile," she told me.

She removed her purple Dora gardening gloves and fearlessly hunted down every escapee bare handed -- letting out squeals the entire time. For Katelyn, rounding up a pound of red wigglers was akin to riding a Six Flags rollercoaster. Her adrenaline pumped as she proudly tucked every last wiggly in for the night. I squirmed just witnessing the act. She was my hero.

A shiver went down my spine as I covered the slimy compost-makers with the remaining bedding and put the lid on. As I tucked myself into bed, I had trouble drifting to sleep as my mind fixated on the worms. *Eew!*

Day 310: The plane roared down the runway toward San Francisco, leaving Orange County behind. Jeff shifted his weight to get comfortable in his seat as his mind raced. The interview went well. Relieved that his composure led to him being able to convey his skills and experience, he closed his eyes and let his head rest against the seat. However, he couldn't relax. What was he doing? A job in Orange County meant finally being able to live near family, but it meant moving away from the environment, friends and lifestyle we loved. A bittersweet day with the potential to change our lives for the better or worse.

Day 311: Sipping iced tea at the backyard picnic table, Leslie and I chatted about Garden Club and she shared her cooking tips with me. Suddenly, Katelyn emerged from her bedroom, tears gently flowing down her cheek.

"What's wrong?" I asked, knowing the signs of a play date gone bad.

"Nothing!" she snapped. "I don't want to play with Megan. She's mean."

"No, she's not. You guys get along great at school," I reassured her.

"I don't want to talk about it," she said crossing her arms and turning away from me.

"We'd better get going anyway," Leslie apologized.

After they departed, I pulled Katelyn on my lap, "What happened?"

"Megan said we're poor because we don't have a TV," she said. "She made fun of me for not having a DS too. And she said my toys are boring and that I don't have enough to play with."

"Oh, sweetie, that's not true. You have plenty of toys, more than enough. And it's our choice not to have a TV. It's not important what or how many things you own. It's about being kind and honest and a good friend. I'm sorry Megan doesn't know that."

Excessive television viewing has been linked to things like autism, violence, ADD, addiction, obesity, sleep deprivation, and even death -- from TVs falling on children.

Reports by Common Sense Media state, "The average child who watches two hours of *cartoons* per day may see more than 10,000 violent acts a year" and "By the time kids enter middle school, they will have seen 8,000 murders and 100,000 more acts of violence on broadcast TV alone."

Maybe Megan was just copying behavior she's seen on TV.

***

Upon hearing Jeff's key in the door that the evening, I rushed to open it and greeted him anxiously to hear about the job interview in Orange County, "Well, how did it go?"

"Good. Really good," he said.

My heart sank. I faked a smile. "I'm so proud of you, " I kissed him.

"Let's talk after Katelyn's asleep," he said.

A couple of hours later, with Katelyn tucked in, I poured us a glass of cool Chardonnay and we curled up by the fire.

"So, you think you'll get an offer?" I asked eagerly.

"It seems hopeful," he admitted. "I was on my game and they seemed impressed with what I had to offer. Is this a move we are prepared to make?"

"There's a lot to think about," I said. "Nine months ago, I would have jumped for joy at the thought. But so much has changed. Unplugging has ruined us," I laughed. "I am so happy, settled and fulfilled now. And Katelyn has absolutely blossomed. I can't imagine not continuing with my newfound interest in teaching environmental education. I will miss the garden, the farm, our friends, the school, our hikes in the woods, our excursions to the city, and our playtime at the beach. It's a tough call," I said, fighting back tears.

"I know, I know. I agree," Jeff replied. "Since Katelyn was born, we've been striving to be near family. I've been on autopilot trying to secure a position in Orange County. Now that we might have one, I'm wondering why we haven't stopped to think about the reality of this move. I'm on the fence. I love it here and have seen the positive changes in our family recently, but do see the benefits of being in

Laguna Beach -- career growth, family support, beach lifestyle, childhood friends."

"Damn. My friends here are so used to me being unplugged that even when we do plug back in, I don't think I'll return to how it was before," I said. "If we move, I'll have to start all over with the awkwardness of being unplugged. I don't know that I'd do it. I really don't want to fall subject to the peer pressure of being plugged back in."

"Let's just see how things progress. I don't have an offer yet," Jeff reassured me.

"I think a trip to Laguna is in order… to do our due diligence and check out real estate, schools, and talk with your family," I said.

"I agree. We need to be prepared to make a decision quickly," Jeff said.

"I'll pull Katelyn out of school and drive down mid-week and you can fly down and meet us on the weekend."

"Deal," Jeff nodded.

As I spooned Jeff comfortably in bed, my mind wouldn't let me enter the land of Nod. Thoughts of life in the OC prevailed. What about my two-year plan -- teaching summer camp at Hidden Villa, a paid Garden Club Coordinator position at Carlson, a one-year Master's in Education degree at UC Santa Cruz. It was such a natural progression for me and felt good to have a plan and be moving forward. Now my momentum would come to a jolting halt. Family comes first. I was relieved at the opportunities for Jeff and Katelyn, but had yet to find a silver lining for myself. *Would this be a decision I would live to regret?*

Day 312: My boobs hurt. *Could I be with child?* Two more weeks until I can take a pregnancy test. A crib, bottles, changing table, maternity clothes…not how I imagined celebrating my fortieth birthday. After our year unplugged, would I revert back to endless hours of Elmo? I honestly don't know how I would've survived my sleep-deprived era without a TV to entertain Katelyn while I catnapped.

Day 313: A day off school for Katelyn meant a Mommy and me day at the Tech Museum of Innovation located only minutes from our house. The museum has all the predictable hands-on exhibits for kids -- build a website, create an Avatar, see how silicon is made, write code. But what I had not considered under the technology umbrella was life sciences, geology, astronomy, and oceanography. The truth of the matter is that advancements in technology have saved lives. At the museum, Katelyn enjoyed being a doctor performing surgery, a scientist growing jellyfish protein, an astronaut exploring black hole dangers, an engineer designing a roller coaster. She even got to use video conferencing to address the "Senate" to give a speech about DNA testing.

Technology is not all about Twitter or Facebook. It's ironic that we use technology to save people's lives, then those people go on to live lives consumed by technology. Upon leaving the museum, we walked through the rain-soaked parking lot, Katelyn taking advantage of the ankle-deep puddles. Running between the raindrops, we ducked inside the dry car as I present a decision-making opportunity for Katelyn.

"Well, should we brave the weather and head to Hidden Villa for our Harvest Feast? It's our last gardening class," I tell her.

Somewhat hesitant, she answered, "I guess so."

With windshield wipers on full speed, we whisked up Highway 280.

Under the protection of an umbrella, Rachel greeted us and directed us to the kitchen where Nancy was assembling the feast. Nancy's hazel eyes looked up at us as she stirred the purple potatoes cooking on the stove.

While waiting for our fellow classmates to join us, Nancy invited Katelyn to adorn her with a butterfly "tattoo" from washable face paints. Katelyn was proud to offer her artistic ability and gently transformed Nancy's arm into a work of art. We soon came to the conclusion that we were the only feast-goers to brave the storm.

"Well, should we go pick some produce for our feast?" Nancy asked.

“Sure, why not? But can I borrow a raincoat?” I said, having left the house ill prepared: no rain boots, no raincoat, no umbrella.

Covered with borrowed ponchos, we trekked out to the garden. Trampling through the saturated garden pathways picking veggies for our feast in the pouring rain, I couldn’t help but giggle. Most rainy days, kids are trapped indoors glued to the TV or entranced in a video game, and here we were getting soaked in search of lettuce, radishes, broccoli, sorrel, and mint. After retrieving our delicacies, we shook off the wetness like a dog after a bath. Our feast also consisted of fresh carrots from our CSA box, apples, oranges, crackers, and potatoes. We nibbled at our organic meal, full of flavor and grown with heart.

Katelyn said to Nancy, carrot in one hand, broccoli in the other, “I hate rain because I get wet, but I love rain because it’s good for the garden.”

With more food then we could finish…Maggie invited the Environmental Education staff to join us. A festive, wholesome meal shared over eco-friendly conversation. It felt warm and comfortable. We left the farm with nasturtiums that had been grown in eggshells that we would transfer to our garden.

In the parking lot, the rain had ceased and Katelyn yelled to the sky, “I love the earth.”

I couldn’t agree more.

Day 317: The Xerox machine buzzed as it duplicated the "Pollinator Power" handout for Garden Club. Another mom entered the teacher's workroom and greeted me warmly.

"Oh, hi, Sharael," she smiled. "Are you making copies for Garden Club?"

"Yes. We're learning about pollinators this week."

"My daughter Frieda is having so much fun at Garden Club. She tells me all about it. How's the fundraising going?"

"Slow, but we're getting there," I admit.

"You know, you should talk to the principal about ordering gardening picture books from Amazon. I work in the school library and we could make a gardening section," she informed me.

"That would be great, but we have no budget for it."

"No, you could use the money we get from Amazon."

"What money?" I ask, with the whir of the copier in the background.

"There's a link we e-mailed to the entire school and if you go through that link to order anything off Amazon, a donation is made to the school. We earn about two thousand dollars a month! That money can then be spent to buy things on Amazon for the school. That's how we bought the classroom reading workbooks. As far as I know, next month's donations haven't been allocated yet," she continues.

"Really. Wow, that would be great. Maybe we could order some garden curriculum books, too," I suggest.

"Yeah, talk to the principal. I do the ordering, so once she approves it I can place the order. Just get her a list of books you'd like."

"That's a wonderful idea, thanks," I said gratefully.

"The other thing is when you send home a flyer for a fundraiser, make sure to add a link so people can make payments through PayPal. The Home and School Club has earned a lot of money that way. Sure, PayPal gets a cut, but it's easier than people having to write checks." And that is the heart of the Internet -- all about making money.

Day 318: Off to the potty again…and again. Oh no, peeing multiple times a days -- not a good sign. When I was pregnant with Katelyn, I knew every public restroom within 20 miles, intimately. One more week until I could take an accurate pregnancy test.

Most women my age and in my predicament wouldn't be opposed to the birth of a second child, and some women spend a lot of time, energy and money for such a miracle. I, on the other hand, am more than a bit gun-shy at the proposition. To me, giving birth would be like going back to a war zone.

Katelyn did not have a joyful entry into this world. Pregnancy, no problem, aside from seeing my body balloon into a size 16 from size 6 and being freaked out at the "alien" moving around in my tummy. Delivery -- blood, pain, screaming…never again.

Day 320: Amidst the fountains, incense, soul-soothing sculptures and books, I was calm, relaxed and at peace browsing through the East-West bookstore in Mountain View…until I glanced at my watch. 2:28 p.m. Yikes! I had to pick up Katelyn from school at 2:40. In my Zen state, I had lost track of time and was at risk for being late to pick up Katelyn…a first. Racing out the door, leaving my peaceful frame of mind behind and focusing intently on the task at hand, *how could I make the 15-minute drive in 10 minutes?* As I pressed the gas pedal aggressively, the clock ticked quickly 2:34. *I'm going to be late. I'm going to be late. Crap.* Katelyn would be left standing alone at her classroom, waiting worriedly. 2:38. *I'm not going to make it.*

Digging through my purse at 68 miles-per-hour, I decided to break the rules and retrieved my cellphone. I could call Sage on her cell phone and have her wait with Katelyn. But what's Sage's cell phone number? I didn't have my address book with me and without using my cell phone, I didn't have new phone numbers programmed into it. I knew she had called me and she had a 510 area code. Could I find it under "missed calls"? Switching my gaze between the road and my phone and back, I frantically pressed buttons -- menus, messaging, address book, instant messaging, calculator, alarm clock. *Where the hell are missed calls?* 2:42. Is it possible that something that had been second nature was now challenging. A year without using my cell phone and I had become illiterate. This could not be happening to me. 2:49. I sped into the parking lot and sprinted towards Katelyn's classroom. There she was, sitting at her desk with the principal teetering on a child-sized chair next to her. Without a word from the principal, I could read her mind, "Terrible mother, late picking up her child! Leaving her all alone."

I thanked the principal profusely and apologized endlessly to Katelyn, who was on the verge of tears.

"Where were you?" she sniffled.

Not having the heart to come clean, I lied, "There was a lot of traffic."

"Well you could've called. I mean this was an emergency. Isn't that what your cell phone is for?" she scolded me.

"You're right." I left it at that and hugged her tightly.

***

"Mrs. Kolberg, I need to talk to you," Katelyn's teacher motioned to me as the children flooded into the classroom to start their day. "We're doing our parent-teacher conference sign-ups online and I heard you don't have access, so we can just do it at my computer here in the classroom after school."

"Great, thanks" I nodded, kissing Katelyn goodbye.

Turning to leave, a fellow mom approached me, handing me a piece of paper.

"Hey Sharael, I know you're not online, so you probably didn't get the e-mail about signing up for conferences online. I printed out the schedule for you. Just let me know what date and time you want, I can sign you up," she said.

I smiled. It *is* possible to live without Internet access. I had created a support system to keep me in the loop and was grateful that I had encircled myself with kind and caring individuals.

Stuffing the schedule in my purse, I said, "That would be great. I'll look it over and get back to you."

Day 321: That's it. I've had it. I can't take this unstable relationship anymore. I give up. I'm kicking my typewriter to the curb. It started off with a warm nostalgia, then turned to a tolerant affection, but once I found out that Sharp has decided to stop making the typewriter ribbon I needed, it's been downhill ever since. Sharp has been making this ribbon for probably thirty years. How ironic that just when I ran out of my supply and wanted to replenish it, they stopped carrying it. *Come on people, I only need a three-month supply! Couldn't you have waited a smidgen longer to stop production?* And I was too cheap to buy a new

$120 typewriter, knowing that it would only be in use for a few months. What were my options? I needed a typewriter, it was not something I could do without. I had agendas to type, letters to mail, book club signs to post. After begging at every office supply store I could find, I came up dry.

Next, I called information and got the 800 number for Sharp. They confirmed that there were no ribbons available for purchase through them, "However, you can find some online through third parties." *Of course I can! Humph.*

Desperate measures were in order. I had heard about a local flea market, surely someone there would at least have an inexpensive typewriter for sale. Katelyn and I piled in the car and headed off. We navigated the rows and rows of makeshift storefronts selling T-shirts, tools and toys. One vendor had dolls that talked on their cell phones. I convinced Katelyn that she really didn't *have* to have one -- although tempting.

Keeping an eagle eye out for an elusive typewriter, we scrounged through tables of used electronic equipment: old TVs, boom boxes, microwaves. After an hour of searching, I found one. A small manual typewriter hidden amongst some old computer keyboards. Not fluent in Spanish, I had inquired at a few booths about a typewriter but was given a blank stare. The term *maquina de escribir* would have come in handy.

I asked the seller, "Cuanto cuesta?"

"Vente dollars.

Cute and portable, I contemplated the purchase. "Quince?"

He thought about it, "No, vente."

On the verge, I worried that had I purchased it, I would once again be the owner of a typewriter without a ribbon. I passed. But the day wasn't a total loss, Katelyn did manage to convince me that she couldn't leave without "one of those beautiful gowns!"

So I coughed up thirty dollars and she left with an off-white sequined quinceanera gown. Just what every six-year-old Caucasian girl needs.

Day 322: I continued my quest for typewriter ribbon, flipping through the Yellow Pages, I stopped at the "typewriters" section and dialed a few numbers. *Bingo!* I struck gold, finding a business that had the typewriter ribbon I was searching for…six boxes!

Located in an industrial part of San Jose, the storefront had a glass case displaying heavy-duty typewriters, the kind they had in the school office when I was a kid. Peering through a door to a back room, I could see stacks and stacks of typewriters, as high as the ceiling. I asked to take a look and was amazed at the amount of inventory.

"You sell a lot of these? Why would someone buy a typewriter these days?" I inquired.

"To address envelopes and labels," the storeowner replied.

My eye caught an antique machine, but when I asked about it, the guy said, "That's not for sale. It's in for repair."

Seventy-five dollars later, I walked out the door and was back in business…or so I thought. Typing away, fully stocked with ribbon, Jeff and I spent a few evenings a night side-by-side in the office, him putting up with the racket of my typewriter, with the occasional comment, "Man, that thing is loud."

Then just as I was typing up the curriculum for Garden Club my typewriter came to a thud upon trying to use my correction ribbon to erase a misspelled word. That was the end of brighter days with my typewriter and me. With the amount of typos I make, not having correction ribbon rendered my typewriter useless. White Out was not an option with drying time and all. That was the last strike.

***

Saturday afternoon, I flung open the front door, stomped to the bedroom and threw myself down on the bed. Jeff and Katelyn were having an afternoon snuggle and story time and didn't dare let my entrance interrupt the princess's dilemma with the pea. I hunkered down under the covers and let out a huff.

"I've had a bad day, " I whined.

Katelyn cocked her head in my direction, "A horrible, terrible, no good, very bad day?"

"Yes," I pouted.

"What happened?" Jeff leaned over.

"I have too much to do and not enough time to get it done and today was a complete waste!"

"How so?"

"I wanted to buy a new heart rate monitor watch since mine got stolen at the gym, so I went to REI and got one. When I got home, while you guys were out riding bikes, I looked in the box and there was no user's manual. It said I could download it online, which I can't! So I had to drive all the way back to REI to return it. Then I drove to Sports Authority, but all they had were ugly black men's watches.

After that, I went to three different stationery stores to find the perfect invitations for our anniversary party and the only options were to pay a bundle for custom cards or to buy the cheaper ones you can print at home, which I can't!

Oh, and I called the credit card company about that fraud alert warning. The charge was from Apple and it didn't go through. It's for my Mobile One account for hosting our family blog. I don't want the site to be taken down so I had to drive down to the Apple store and pay in person. Then the sales clerk handed me a software box and said, 'Here's the code. Just enter it online and you're good to go,' which I can't! She was nice enough to do it for me there in the store when I told her I don't have Internet or a computer. She must have thought I was a wacko wanting to purchase software without those things.

Anyway, I'm so frustrated! I could've bought a watch online, ordered cheap invitations online and paid for my Apple software online, all in less than an hour! Instead, I wasted my entire day running around fruitlessly," I paused, with a pout…"And I'm getting fat!"

"Oh, Mama," Katelyn said, reaching over and rubbing my back. "I know what it's like to have a bad day. You know at school we have so much work to do and never enough time. The other day we were doing this really cool art project and we wanted to finish, but the teacher took it away and shoved it in our cubbies because we had to move onto the next thing. She said we could finish it at the end of the day, but we never got to. But don't worry, Mama, tomorrow's a new day. I'm sure it'll go better."

How did my six-year-old become so sweet and wise? Her words and snuggles melted my worries like a snowman on a hot day.

"Thanks, sweetie, I needed that" I kissed her.

Jeff reached over Katelyn and softly stroked my cheek.

"I didn't think I'd feel this way, but it'll be a relief to be plugged in again," I said, resting my head on the pillow.

Day 323: The chickens clucked quietly as I strategically chased them down, grasping at air as each one escaped my grip. A handful of chickenfeed and they came flocking to me. The kindergarteners giggled with glee at my attempt to catch a hen. Finally, a distracted subject, pecking at the feed on the ground was mine. Approaching from behind, I lunged at her, keeping her wings tight against her body. Slowly, I stood up and pressed her comfortably against my torso, her feet tucked in tight.

"Kids… would you like to pet the chicken?" I called out. "She's very soft. Don't worry, she won't bite, she's very friendly."

The chickens at Hidden Villa had grown on me. The hen felt warm against my body as I coaxed the kids to gently feel her feathers. I was flying solo. My first day as an official Farm Guide. My job description included leading a group of 10 kinders around the farm with stops in the garden for taste testing, and up-close-and-personal encounters with pigs, sheep, goats, cows and chickens, all the while encouraging a respect and curiosity for the natural world and promoting an eagerness to protect our environment for the future.

"Who wants to meet our piggies?" I asked the kids.

"I do!" they yelled out with raised hands.

The chickens parted like the Red Sea as we continued our tour. My heart soared at doing something that felt right.

***

Over dinner, I told Jeff about the day's activities. "Oh, and you'll never believe it, Sage's husband Steve went to the dump today and invited me to go with him…but I had to work at Hidden Villa. Can you believe it? I was so bummed."

"Oh yeah, big bummer, you didn't get to go to the smelly dump," Katelyn shook her head.

"And he said they even give tours! I'm telling you, you guys are missing out," I continued.

After dinner, while doing the dishes, I reached over to the windowsill and removed a seedling from an egg carton turned painter box and held it up for Jeff to see,

"Look, I planted some Nasturtiums in eggshells," I showed him.

"OK, this is getting out of hand," Jeff laughed. "Inviting neighbors to pull weeks, planning play dates at the dump, now eggshell flower pots! You're too much!"

I giggled, "Oh, it's become so second nature now."

Day 324: Katelyn and I are road tripping…off to Laguna Beach to investigate schools and real estate in preparation for Jeff's potential job offer. Since it's just the two of us, I thought I'd break up the drive by doing an overnighter along the way…but where?

"I can kill two birds with one stone…keep my promise to the man who helped us when we ran out of gas, and get some travel tips…let's go to AAA," I said to Katelyn

"But Mama, why would you want to kill birds? I thought you loved animals," she said concerned.

"Oh, my love, it's just a saying meaning that I can get two things done at once," I told her.

So off we went to the local "Auto Club" as my in-laws call it. The office was rather lively with people lined up to get help with their travel needs. Finally, our turn. We saddled up to the counter covered with a map of the US. With my shiny new AAA card in hand. I inquired, "We're new members. What does our membership get us?"

A middle-aged woman with long brown hair and large glasses answered with an Hispanic accent, "Oh, lots of wonderful thing…maps, directions, hotel reservations, tickets to Disneyland."

"Great. We'd like to go from San Jose to Laguna Beach with an overnight stop somewhere in the middle," I informed her.

She turned around and reached into a large filing cabinet and retrieved maps -- Bay Area, Central California, Orange County, Los Angeles, Santa Barbara. She unfolded and spread out the California map. Katelyn raised on her tippy toes to peer over the counter at the map. As I ran my finger along the highway routes between San Jose and Laguna Beach, I stopped on a point of interest I thought Katelyn would get a kick out of...Solvang. A Dutch town. It would be our faux European vacation.

“That’s it. We’ll do a night in Solvang. Now, how do I get there?”

The AAA agent reached for a highlighter and circled San Jose with her fluorescent yellow pen, then Solvang, then Laguna Beach. Then she grabbed her orange highlighter and marked the route to take – Highway 101 South to Solvang. Easy enough. From Solvang to Laguna Beach -- not as easy. We studied the map closely.

“You could just take Hwy 1 from Santa Barbara to Laguna Beach,” she said looking over the map.

“Oh, I think that would take forever,” I replied.

“How about continuing on the 101 or going inland to 5,” she said.

Her lack of confidence in suggesting the quickest route did not sit well with me.

“Do you have MapQuest or something that might tell the best way to go?”

A bit flustered, she replied, "Sure. Let me punch it into our computer. Here's a complimentary guidebook you can look over while you're waiting."

She flipped to the Solvang page that offered a two-paragraph description and recommended hotels and restaurants. Katelyn and I paged through the book while waiting patiently. Nearly 10 minutes later our agent apologized, "The directions are still printing. There are 11 pages."

She returned, a thick stack of directions in hand, and then proceeded to read each page and transfer them onto my map via her trusty highlighter. Forty-five minutes later, we knew how to get to our destination.

"So, I guess I'll go ahead and make a hotel reservation while I'm here too," I said.

"Great. Louise can help you with that," she said, motioning for us to wait in an adjacent line.

Thankfully Katelyn had gone potty and filled up on a Jamba Juice prior to us stepping foot into AAA. After Louise helped some folks in front of us book hotels from Fresno to Mexico, it was our turn.

"Hi. I'd like to book a hotel room in Solvang please," I said, excited about our trip.

Looking gleefully at Katelyn she smiled and said, "Come on back to my desk and get comfortable."

Her workspace was scattered with travel brochures.

"So, you're going to Solvang? One of my favorite towns," she said, looking at us over the rim of her reading glasses. "I'm part Dutch, so it's an obligation for me to visit there frequently. I highly recommend this quaint little hotel nearby Solvang where your daughter can ride a mini-pony. It's an adorable little place."

A sweet grandmotherly woman, I felt almost obliged to accept any recommendation for fear of hurting her feelings if I didn't, but I spoke up.

"We'd actually like to be in town, so we can walk to dinner. It's just a quick stop."

"Is it just the two of you traveling? And how, much do you want to spend?" she asked.

"Yes, just us. About a hundred dollars, I guess."

"Well, let's see what's available," she said, tapping away on her computer. "Oh, a great place right in town, next door to the best Danish bakery, for ninety dollars. I've stayed there myself. You'll love it."

I cranked my neck to glance at her screen.

"Do you have a photo of it?"

"No, but it's a traditional Dutch style. Very charming. And clean," she said convincingly.

"We'll take it," I said.

She typed our request and my credit card info into her computer. Moments later she was agitated, "Hmm. It wants to charge me for three guests when clearly I typed in two. Let me just close this out and start over."

Second time was not a charm.

"Well, I suppose I could just call them," she finally said.

The thought had crossed my mind. *Couldn't I have done that myself and saved a lot of time?*

On the phone she said, "Hi, this is Louise from AAA. I need to book a room for Wednesday, two people, non-smoking, bottom floor. There's a little one, so it would be easier not to drag the luggage up the stairs. I've stayed at your hotel myself and loved it, I'm sure they'll be quite happy here. What's your rate? Oh, eighty dollars. We'll take it," she said winking at me.

Hanging up the receiver she said, "Sometimes it's better just to call and have the person-to-person interaction. We got a better rate after all. So let me just type up this confirmation for you."

I would've been just as happy with the handwritten form she had filled out. While she was typing, I commented, "With sites like Travelocity and Yahoo Maps, I'm surprised the Internet hasn't put you out of business."

"Oh, we stay very busy here," she replied. "People want to talk to someone that's been where they're going. I've been doing this for 25 years. Our customers enjoy the advice and recommendations I can give them, and the discounts. We get a lot of people wanting to go to Disneyland. That's a big business for us."

After handing me my hotel confirmation, she said to Katelyn, "Which Disneyland characters do you like best – Mickey, Cinderella?"

Katelyn shrugged.

"All of them, I think," I said.

Louise reached beneath her desk saying, "You've been such a good girl. I keep a stash of prizes for good kids that come in here."

She popped up handing Katelyn a bracelet with a plush Cinderella head on it. My first reaction was that Katelyn was going to be ungrateful and find it too "babyish," but Katelyn accepted it with delight.

I replied, "You couldn't get that by booking a trip online!"

As we got up to leave, Louise commented on Katelyn's sparkly gold Converse shoes. Katelyn smiled with her whole body as if it was one of the best compliments she's ever received. We left the office ninety minutes after we had arrived, with all we needed for a joyous journey ahead.

As I retold our story over dinner to Jeff, I grabbed our pile of travel materials -- four maps, a guidebook, 11-pages of directions, and a hotel confirmation.

"I could bring all this, or just use an iPhone!" I said to Jeff.

It did seem like a waste of paper -- and time.

Then Katelyn added, "But I wouldn't have gotten this cool bracelet and we wouldn't have met Louise. She was so nice. I love triple A."

***

Perched on the potty, my hand between my legs clutching a plastic stick that would foretell my fate, my stomach churned with anticipation of the results. I stood alone in the bathroom -- the house silent -- as I waited for the results of the pregnancy test. Longest two minutes of my life. *One line -- not pregnant, right? Oh, God.* I reached for the package to make sure I was reading the stick right. *Not pregnant. Thank God,* I whispered, feeling selfish, but relieved that my life could go on seamlessly. Katelyn would not get that little brother she'd been wishing for. Actually, she wished for an older sibling, so that wouldn't have happened anyway.

Day 325: We rolled into Solvang in the late afternoon and checked into our AAA-recommended hotel -- a Best Western in the heart of Danish country. Unloading our suitcases, we walked the streets in search of an authentic smorgasbord dinner. A couple blocks away we stumbled upon a farmer's market and eyed the produce, settling on a basket of strawberries. Shops were closing for the night, so we peered through store windows -- wooden shoes, garden gnomes, and Danish pastries. A large wooden windmill rose above the town in an attempt to solidify the feeling of Denmark.

Upon recommendation from our AAA guidebook, we opted for dinner at the "Bit O' Denmark." We slid into our faux leather booth as the waitress pointed out the traditional smorgasbord that included head cheese (meat from a pig's head), pickled herring, Spegepølse (salami), fish frikadeller (fish meatballs), and pork roast. Being a vegetarian, the Danish fair didn't quite appeal to me. So, I selected the cheese platter. Katelyn had an off-the-menu grilled cheese sandwich. Choking down my pungent cheese, I asked the waitress, "So, can you tell us a bit about Denmark?"

The woman wrinkled her weathered face and tossed her red, curly bob and said, "Never been there. I don't know anything about it."

Traitor, I thought. *How can you work at the "Bit O' Denmark" and not know any interesting facts to satisfy tourists curiosity?* Seems like it would be required during training.

"There is another woman who works here that's visited Denmark, but she's off tonight."

Certainly, I could've looked up some interesting facts on my iPhone to tell Katelyn over dinner. After sending back my cheese plate, I wolfed down Katelyn's fries and we ordered dessert. The Abelskivers -- Danish pancakes drenched in strawberry sauce. A huge hit. Back at the hotel we were spoiled, flipping TV channels between *Snow White* and *American Idol*. Staying up way past bedtime, Katelyn and I snuggled by the glow of the TV that illuminated the dank hotel room.

Day 326: Glancing at my watch, the time was 8:11 p.m. I jumped up from the dining room table that was strewn with Katelyn's homework.

"*American Idol* is on!" I exclaimed.

We rushed to the living room. My father-in-law clicked on the TV as we gathered around and Jeff and I acted as sportscasters giving a play-by-play account of what was taking place. As *American Idol* virgins, Jeff's parents were stunned at the level of talent on the show and the raw emotions of getting voted off. During a muted commercial break, my father-in-law noted, "I'll finally be able to follow what my friends are talking about at breakfast tomorrow."

Pop culture is something my in-laws have never been devoted to, which at times, leaves them conversationless when the topic comes up. Yet, the reality is that keeping up with pop culture is only a click away. As the show progressed, my mother-in-law slumped in feigned interest, up past her bedtime. Katelyn took on the role of the fifth judge, quickly dishing out her professional opinion on the contestants' performances. At the end of the show, my father-in-law asked, "So, who won?"

Katelyn explained the rules and I interjected, "Want me to set your DVR to record it from now on? Then you can watch it whenever you want and fast forward the commercials."

"You can do that? Does our TV have that?"

"Yeah, let's set it up tomorrow," I offered.

Day 327: The leather-seated Lexus sped down Pacific Coast Highway, Katelyn and I buckled in the back and Jeff in the passenger seat with Seth behind the wheel. Our chauffer for the day, Seth's job was to find us the home of our dreams in Laguna Beach for less than a million -- a difficult task to say the least.

My in-laws put the pressure on Seth as he escorted us out the door of their house.

"It's up to you Seth. Find them a house they can't refuse," they told him, with hopes that we'd make Laguna Beach our home.

"We're just looking, Pop," Jeff reiterated.

It just so happened, a house was on the market just across the street. A two-bedroom, one-bath with a small yard, and killer ocean views for only eight hundred thousand, a bit overpriced for the condition of the property.

The waves lapped at the shoreline as we careened down the highway towards the next option. Seth flipped through his MLS printouts he had researched online. Each house had basically the same stats, just different floor plan. By the end of the day, I felt discouraged by our options. Our little corner of Northern California suddenly seemed like a bargain.

Day 329: Katelyn ran to me and wrapped her arms around me when I stepped into her classroom to pick her up.

"Well, hello. How was school?" I asked.

"Great, we danced to a *Hannah Montana* song for dance class."

As she gathered her belongings and stuffed them into her backpack, Mrs. Nelson approached me.

"Do you have time to sign up for conferences now?"

"Oh! I totally forgot. Teresa was going to do it for me, but with Katelyn home sick the last couple of days, I didn't get back to her. Sure, let's do it now."

She jiggled the mouse to snap her monitor out of sleep mode. Her antiquated blueberry iMac sat peaceful on her classroom desk.

"Do you know how to get to the website?"

"Uh, no."

She typed in the URL and pulled up the page.

"So you just choose the date and time and type in your name next to it. You can scroll down for more dates. Do you know how to use this?" she smirked, her hand on the mouse.

I chuckled, "Yes, I do know how to use a computer, we just don't have online access at the moment."

Day 330: My first organized book club. I had the questions we'd discuss. The time and place was set. I call to invite potential members. Friends should be arriving any minute to have an intellectual debate about literature. Did I sit at the head of the table waiting anxiously for them to arrive? No. Instead, I sat in the front seat of my Prius, halfway between Orange County and the Bay Area -- hours away from my infant book club. I would miss my first gathering. The unfortunate thing was that I did not have the access to send a group e-mail to cancel the event and had gotten no definitive answers to my invites. I waited for my cell phone to ring, thinking whoever attended might attempt to reach me on my cell phone to inquire about my whereabouts. It never rang. A tribute to my efforts to inform my friends that I don't use my cell phone anymore. *Did anyone show up? Would anyone be irritated with me for inviting them and not being there?* I might never know.

Day 331: Reluctantly I stepped on the scale…135! I let out a deep sigh. Our unplugged lifestyle has taken a toll on my waistline. The 15 pounds I lost last spring have crept back on, plus an extra five.

The argument has been widely publicized that to be fit and healthy it is best to turn off the TV. In my case however, during the cold, rainy days of winter I would normally welcome the TV as a way to be entertained while running on the treadmill. Without it, I was less motivated to drag myself to the gym.

While working in the high-tech industry in San Francisco, I remember a colleague of mine had her cubicle desk raised so that she could stand while working -- saying that she would burn more calories that way. She would have loved the TrekDesk that actually lets you turn your treadmill into a workstation and allows you to walk off calories while getting your job done. *Why doesn't every company offer that as a preventative health measure?*

On sunny days, although I enjoy embracing the sounds of nature, I find it easier to skip runs without the lure of my iPod to keep my company. Apparently, I'm not the only one that feels that way. *Body+Soul* magazine reported, "The right music can make you look forward to exercising, and it might even help you build stamina."

So, unplugging led to me gaining 20 pounds. It couldn't have possibly been those extra servings of Ben & Jerry's ice cream.

Month Twelve: To Be Online or Not to Be Online?

"All this modern technology just makes people try to do everything at once."
--David Watterson, Calvin and Hobbes author and cartoonist

Day 333: Three weeks ago, I got what seemed like a simple request, "Can you get me a photo?"

Sage had found someone to donate a picnic table for the school garden, so the kids would have a place to do their lesson plans. The principal needed a picture of the table to show the school board, to get it approved. On the day of the request, I tracked down the donor, asking, "Can you e-mail the principal a photo of the table?"

"Oh, I don't know if I know how to get it onto my computer," the older lady admitted. "My husband probably can figure it out. Also, I don't have the principal's address."

"You can just send it to Sage," I told her over the phone.

A few days later, the principal approached me, "How's the photo coming along?"

"Oh, didn't Sage e-mail it to you?"

"Not yet."

Upon talking with Sage, it had been sent, but whether it was the correct e-mail, was uncertain. Apparently it didn't go through.

"I'll just have Steve print it at Target and I can go pick it up," Sage said.

A few days later, I ran into the principal.

"How's the photo coming along?"

"Didn't Sage drop it off to you?"

"Not yet."

Later Sage told me, "Yeah, Steve got too busy and I never made it to Target anyway. I'm so sorry."

"Oh, no big deal. I'll just take a photo myself and have it printed."

So, I spent the next few days snapping shots to use up most of my roll of film. I drove to the donor's house, snapped the shot and high-tailed it to Target. I rewound the film and popped it out of the camera, ready to drop it off at one-hour…then, I saw something on the canister that put a dent in my plan, "Kodak Professional Film." Those three words meant it could only be developed at a professional lab, thirty minutes away in Palo Alto. *Ugh.* The next day was spent lollygagging around Palo Alto, waiting five hours for the roll to be printed then driving the 30 minutes back home.

Finally, print in hand, I delivered it to the principal. She looked it over and casually replied, "Oh, this will never get approved by the district. It's wooden. Kids could get splinters. Skateboarders could use it as a ramp on the weekends and get injured. Too much maintenance and liability. You'll need a metal table.'"

Three weeks of effort down the drain.

"Just write a letter to the Home and School Club and see if they can pay to order a table through the district."

Somehow I didn't think it would be that simple.

Day 349: The Easter bunny laid tie-dyed glitter-covered, crayon-decorated eggs strategically throughout our house and into the rain-sodden yard. Katelyn sprung out of bed, basket in hand, ready for the hunt. With no competition, she hopped from egg to egg, plunking them in her basket. My Easter's as a child were not so merry. With a sister five years older, I didn't stand a chance. She'd zoom through the house and pluck the eggs from their hiding spots while I shuffled along slowly in my footy pajamas. The morning usually ended in tears. I cried because I didn't find the eggs, my sister cried because my Dad took the eggs from her basket and game some to me.

By seven, Katelyn had devoured Peter Rabbit's chocolate ears and half of his face, a handful of Jolly Rancher jellybeans, a half-dozen foil covered chocolate eggs and an unknown amount of Whopper eggs.

Then came breakfast. Fresh fruit and scrambled egg whites, right? Nope. Pancakes doused in maple syrup, of course.

By mid-morning we headed off to church, conveniently located across the street from our house. Our obligatory Easter service attendance. Katelyn pranced over, her pastel, floral-print Easter dress flowing like the butterflies around us. Upon entering the church, we were offered a newcomers coffee mug. I guess attending twice a year for three years didn't make a lasting impression on the congregation.

As we sat in the pews, the choir belted out seasonal hymns while Katelyn listened intently to the angelic lyrics. The sermon followed…"died…death…dying." Katelyn covered her ears in protest. I had forgotten about the "Jesus dying" part of Easter and had only remembered that he rose to heaven. As the sermon continued, suddenly the focus of Easter seemed very grim, with only a closing paragraph about the resurrection. This did not enlighten our Easter morning.

An elderly woman invited the children up to the front for a lesson that went something like this, "At Easter we like to talk about metamorphosis -- how the caterpillar spins a *web* into a cocoon and then *pecks* its way out and turns into a butterfly -- much like how Jesus changed from human to spirit."

Well-intended meaning, poorly researched execution. The day went downhill from there. Easter afternoon, Katelyn amped on sugar, pouring rain, everything closed. So much for our Easter candy-burning hike. No stroll in nature, no bowling, no rock climbing, no park, no roller-skating.

"I've got it. How about a movie?" I spewed.

"Yeah, *Diary of a Wimpy Kid* is playing," Katelyn said.

"We've already seen it," I reminded her.

"I don't want to spend money on a movie you've already seen," Jeff said.

Had we known about the two-hour wrath that was about to ensue after our comments, we would've willingly agreed to go to the movie.

Katelyn burst out with, "But I really want to see that movie. It's *sooo* good!"

Fast forward to an hour later and several alternative options that were shot down by my movie-loving six-year-old.

"I hate money. I wish money wasn't in life. I just wish everything was free," Katelyn sobbed, face down on her bed.

Jeff and I formulated a peace treaty.

"How about we go, if you pay for your ticket with your allowance," Jeff says.

Katelyn sat up, her face red and blotchy, her eyes wet with tears, "That's fine! I don't mind using my money. Let's go."

If only solving all world peace issues were that easy. But they're not. Neither was our attempt.

"Bad news…the movie already started," I said. "And the next one's not on until after bedtime."

Katelyn screamed uncontrollably, "I don't care if it's on at one in the morning, I want to go see that movie!"

Her severe reaction startled us.

"Life is so unfair. I hate this world," she cried.

I grabbed her and pulled her onto my lap, stroking her blonde curls, my eyes welling up, "What's really going on? Are you upset about us possibly moving to Laguna?"

She looked at me with bloodshot eyes, "No, it's not that. It's just tough being a kid. I didn't understand what church was about. I don't know how much things cost -- except for a penny, I'm not that good at riding my bike yet, I don't always get to decide how we spend our time. It's just hard."

She buried her head in my shoulder.

"Oh, little love. I know. I was a kid once too. Things aren't always easy," I said softly. "But how about we put this day behind us and focus on the things we have to look forward to this week instead -- swim lessons, garden party, recital."

She nodded in agreement and added, "OK, but I really wanted to see that movie."

"I didn't realize it was that important to you," Jeff said putting his hand on her back.

"It's just that without a TV, I really like to go to the movies. It's been hard. I just can't wait two more weeks to get the TV back," she whimpered.

"You've been so patient. I know it's not easy. You've been so good about not having a TV," I said.

Day 352: Flipping through the pages of *Time* magazine while burning off a snicker doodle on the elliptical, I came across an article by Joel Sten. He discussed his failed attempt to unplug for 24-hours. Apparently, there is a National Day of Unplugging, but I hadn't heard about it because I'm unplugged. So inspiring to hear people are at least contemplating the thought of trying a life without technology.

Day 362: It all started with a dare. "I can't wait to get my laptop back. I'm so tired of being cooped up in my home office because of my typewriter," I said to Jeff. "I miss being mobile with my laptop, like taking it to the Roasting Company and enjoying the scene while working."

"Why don't you take your typewriter to the Roasting Company? They have outlets," Jeff joked.

"Very funny," I said, sticking out my tongue at him.

"I'm serious," he smirked. "Although, they might kick you out because it's so loud. I *dare* you to do it, but on a weekend, so I can watch…and laugh."

"Are you kidding? That would be so embarrassing. I couldn't do that -- walk into a coffee shop full of laptop-toting entrepreneurs with my '80s typewriter. They would think I was such a dork, or completely insane. Los Gatos is a small town. People would talk," I said.

"Maybe you'd start a new trend. You just gotta do it," Jeff egged me on.

So, here we are, the Sunday before we begin plugging back in and I decided to take Jeff up on the dare.

"OK, I'll make you a bet," I bargained. "If I do this, you make dinner for a month."

"Sure, no sweat," he said, much to my shock.

"…and, do the grocery shopping once a week," I added.

"It'll be worth it," he said grinning widely.

Jeff and Katelyn walked in ahead of me, ordered waffles and took a seat at the coffee bar.

As I sauntered down the sidewalk, my palms started to sweat when I neared the Roasting Company. My typewriter hung heavy in my hand. Holding onto its convenient portable handle, I tried to act nonchalant. Sunday morning in downtown Los Gatos is a social mecca -- the farmer's market is buzzing, the town square is full of lively families, dog lovers are strolling the streets with their pooches, cyclists stop to chat and refuel with caffeine. The Roasting Company was hopping. The window seats and benches outside were full with cyclists, dogs, and coffee addicts. I strolled up, opened the shop door, took a deep breath and scanned the crowded shop for a seat, next to an outlet.

I saddled up to the bar, at the opposite end from Jeff and Katelyn, and plunked my typewriter down next to a man who was simultaneously reading the newspaper and texting on his phone. I thought for sure he would gasp at my contraption. Not a peep. I plugged in, pulled out my clean, white typing paper and slung my jacket over the back of the barstool. Trying not to look nervous or out of place, I got in the queue for coffee. In a room full of iPods, iPhones, iPads, and laptops -- my typewriter stuck out like a zit on the end of a teenager's nose!

A cup of steamed milk calmed me as I was about to really make a scene. The smooth white paper rolled easily into the typewriter with a twist of the knob. Then, *tap-tap, tap-tap-tap-tap*. Each time the key struck, I grimaced with embarrassment. *I am not a dork. I am not a dork,* I reassured myself. *Jeff better make some damn good dinners!*

My fingers rigidly stroked the keys. Luckily, with the chatter of the full house, the espresso machine working overtime and the coffee grinder hard at work, my typewriter was not as noticeably loud as I had imagined it would be. I set to work on a document I needed to complete to get additional funding for the school garden.

After a few moments, I began to think, *oh, good, no one even cares. They are too wrapped up in their busy lives to notice my typewriter and me. Maybe it would be uneventful.* Just then, an old man approached me, "Is that a typewriter?" he practically yelled.

"Yes, it is," I said as my face flushed.

"I'm hard of hearing, but I thought I heard a typewriter, so I had to come see it. They still make those?" he asked.

"Yes, they do actually. But this one is from when I was in college. My dad found it in his garage and dusted it off for me. I just thought I'd give my laptop a break," I replied.

"That's just swell," he smiled jovially. "It must feel nice to type with that instead of a computer."

"Yeah, it's fun. Wanna give it a try?" I offered. By this time, several sets of eyes were peering at us from behind laptop screens and coffee mugs.

"Oh no," he chuckled. "I'm not very good at typing. I only took one class in high school. That was many years ago."

"Do you own a computer?" I asked.

"Yeah, I use the 'hunt and peek' method. It works for me," he shrugged.

Just then Jeff leaned over and, trying to add to my embarrassment, said, "Excuse me, Miss. Can I take your photo?"

He raised up our 35 mm camera and the old man yelled, "Hey, that looks like my old Pentax camera…that *is* my old Pentax camera. Well, what do you know."

Between the camera and typewriter, he must've felt like stepping back in time to his glory days. He went on to tell me about a typewriter symphony that he'd heard, "You can look it up online."

Our conversation ended and he patted me on the back and took a seat at his table and I overheard one of his friends comment, "She's living in the dinosaur era." Obviously, the senior citizens are hipper than me and my unplugged way of life. After the excitement, Jeff and Katelyn hit the farmer's market while I stayed behind to finish my document, now comfortable with my pseudo celebrity status.

Not may keystrokes later, I was out of typewriter ribbon, so I decided to close up shop. As I headed for the door, a man seated at one of the tables stopped me and peering through his think-rimmed glasses, said, "What's the deal with the typewriter?"

"Oh, my family and I are taking a break from technology and I'm tired of being stuck at home, so I thought I'd get out and bring my work with me," I explained.

"Oh, everyone's wondering what you're doing," he said, scratching the scruff of a beard on his face. "We thought maybe it was a psychology experiment. "

"It's nice to get a break from my laptop. You should try it," I encouraged.

I left hoping that just maybe I had inspired someone to contemplate their overuse of technology and realize they do have alternatives. As I crossed the street a cyclist zoomed by and shouted, "Hey, is that a typewriter?"

"Yep! You should get one," I shouted back.

Maybe Jeff was right. Maybe this could catch on. Maybe it would just take one brave (albeit reluctant) soul to lead the way. Maybe I shouldn't be embarrassed, but proud. Maybe I would have to do this again -- to reach more people. Jesus' prophets had the bible. I have my typewriter -- spreading the word of an unplugged lifestyle.

Day 363: Finally, a call came through that I had been waiting months to receive.

"Hey Sharael. It's Sam Thomas calling from the Michael Lee Foundation. I have good news. I met with the board this week and we've decided to grant your request for the fifteen hundred dollars for the Carlson Garden Club."

"You're kidding! That's amazing. Thank you so much!"

This is a success that has been a true testament to the worthiness of living unplugged.

It started when shortly after joining the Sierra Club, Jeff and I received the newsletter in the mail. In the newsletter was an ad for an environmental fair. There was a website and phone number listed with the date of the fair, but no address. It was a 408 area code so I knew it was in the area and wanted to attend. Without Internet access, I couldn't visit the website to get the address, so I picked up the phone and dialed.

A friendly woman's voice answered.

"Hi. I'm calling to find out were the environmental fair will be held," I said.

"Oh, thanks for calling! It's gong to take place at Mount Madonna School in Morgan Hill."

"Oh, at a school. That's nice," I said.

"Yeah, and if you have kids bring them along. We will have lots of crafts and activities, and they have a school garden the kids can plant in."

"That's great," I said. "I run the Garden Club at my daughter's school. The kids love it, but we're having trouble because we have no funding."

"You should call the Michael Lee Foundation. They helped us get our garden up and running. They're a non-profit that funds environmental projects in the area," she informed me.

"That sounds great. How can I reach them?"

"They have a terrific website with all their info listed."

"You don't, by chance, have a phone number do you?"

"Let me look it up for you," she obliged.

Hanging up, I immediately dialed the number. A man answered. The connection was bad and I could barely hear him.

"I don't know if I have the right number," I said. "I'm calling about getting funding for a school garden."

"Yeah, we can help with that," his muffled voice said.

As I told him our story, I struggled to hear his responses through his Blackberry.

A few weeks later, he was in our school garden telling me the story of Michael Lee -- a successful high-tech entrepreneur who died young in a surfing accident and left a legacy to promote environmental education in Santa Clara.

I welcomed him to our humble outdoor classroom and invited him to watch our Garden Club in action. During lunch break, kids lined up on cue to toil in the garden. Leslie, my co-volunteer, took half the group, about 15 kids first through fifth grade, and weeded, watered and planted in the planter boxes. I took the other half and led them through a lesson plan about the parts of a plant. Sam listened, took notes, snapped photos, and offered suggestions.

After the whirlwind of activity ended by the ringing of the school bell, Sam said to Leslie, and me "I'm very impressed with what you've accomplished with only a two hundred dollars and some small donations. The kids are really engaged."

"As you can see, they love it," I said, my hands dirty with soil. "Not only are they learning about gardening, but also about teamwork, friendship, and environmental stewardship. And utilizing the space to reinforce State Standards has been well received by the kids, the principal and the staff. The students learn math, reading, writing, science, and art -- in the garden."

We walked over to the principal's office to meet with her and fill her in on our plans to create a sustainable garden program. During our discussions, the principal turned to Sam and said, "I haven't' talked with Mrs. Kolberg about this, but would your organization be able to provide funding to pay Sharael to run the garden? It would be great to have her as a resource so the teachers could utilize the garden and incorporate it into their curriculum."

I was so flattered and excited about the new job prospect.

"I'm sure we could work that out somehow," he replied confidently. "Sharael has been so professional, passionate and persistent in trying to obtain a grant for your school. I'm sure we'll be able to provide some sort of financing. And, I want you to know, Sharael, how much I appreciated the handwritten thank you card. You don't get those very often these days. Those personal touches make a big impression."

Upon our initial conversation, I had mentioned to Sam that I did not have access to e-mail or Internet. At that time, he told me, "I respect that. I can see the benefits of living that way."

Shortly after our meeting, I sat down at my typewriter and put together a 10-page proposal for the garden grant. I had never written a grant proposal before and certainly not on a typewriter. But it worked. I stuck it in the mail and awaited the call. It seemed as thought the funding was a sure bet, until a couple of weeks ago when I hit a snag.

I was bragging about how we were going to get a big check for the garden without having used e-mail, Internet, laptop, word, and digital photos. Quite an accomplishment!

Then Sam called, "I need you to e-mail me the proposal so that I can get it to the board before our meeting in a couple of days. And can you just make it a summary. I tried to scan it in, but our scanner broke."

"Well, I don't have e-mail, but I'll see what I can do," I admitted.

I rushed over to Sage's, since she *does* have e-mail.

"We're so close to getting the funding, but he needs an electronic copy. Do you mind if I dictate it to you?"

She agreed without hesitation. After about an hour, the shorter, electronic version of the proposal was sent into cyberspace Sam called to congratulate me, he said, "I know you went out of your way to get that e-mail to us, but it really made the difference."

After he congratulated me for receiving the grant, I said, "I'm so excited and the money's going to make such a big impact, but I do have some bad news. My husband just accepted a job in Orange County -- we're moving this summer, so I won't be here next year to run the garden."

"That's more than a little disappointing," he said.

"Tell me about it! But I will make sure to find someone to take over for me before this year is up. We do have a sustainable plan in place, so I am confident that the garden will flourish, especially with the funding," I said.

"Let's meet to discuss how to move forward," he replied.

"I'm willing to stay involved if there's a chance for me to do so from Orange County. I've seen first-hand how important it is to provide kids with this opportunity and would really like to ensure the money is spent properly. I was looking forward to working with you to roll this program out district wide," I told him.

"Let's discuss how we can best utilize your service. I have no doubt that once you settle in Orange County opportunities for this type of involvement down there will sprout."

"Maybe I could write a garden blueprint that can be utilized for any school to start a garden? We could put it on your site as a PDF download," I offered.

"Actually, the web has tons of resources like that available," he said.

Of course it does, but I wouldn't know since my gardening expedition started after we unplugged.

Day 365: I can't believe it's been an entire year since we unplugged. Tomorrow we return to the mainstream – well, at least we have that option. I am both excited and sad. Relieved to resume technology use for convenience, disappointed to no longer live a life of simplistic communication and family bonding. I worry that once we cross back over that line, we'll get sucked back in and our year of living off the grid will become a faded memory in an instant. Yes, we've had frustrations, but we've seen our family grow and encountered opportunities we would not have had if we wouldn't have been brave enough to unplug. Katelyn has a new appreciation for simple pleasures, respect for the environment, and fascination with things from the old days. Jeff was awakened to the physical, mental and emotional impact of useless TV viewing, as well as an even greater interest in music.

My life has undergone a metamorphosis from web producer and online journalist to earth lover and solitude seeker. I hope through our experiment, our friends and family have learned how technology can impede deeper relationships. The biggest impact has been that we have learned to be independent thinkers and cherish our uninterrupted family time. Just because everyone texts, e-mails, watches TV, buys/sells on Craigslist, surfs the Internet, shoots unlimited digital photos -- doesn't mean we have to. There are options. I feel grateful to have that newfound awareness -- to know I don't have to go with the flow, but can contemplate alternative ways to live my life and by doing so become a more well rounded, grounded and open-minded human being.

Resources

Books:

Blakey, Nancy. *Go Outside: Over 130 Activities for Outdoor Adventures*. Berkeley, CA: Tricycle Press, 2002.

Dutwin, David. *Unplug Your Kids: A Parent's Guide to Raising Happy, Active and Well-Adjusted Children in the Digital Age*. Avon, MA: Adams Media, 2009.

Gil, E. *The Healing Power of Play*. New York: Guilford Press, 1991.

Hicks, Mack R. T*he Digital Pandemic*. Far Hills, NJ: New Horizon Press, 2010.

Israel, Shel. *Twitterville*. New York, NY: Penguin Group, 2009.

Kanten Hartfield, Wanda. *Unplug! 101 Ways to Pull Your Kids Away from Television*. Trafford, 2003.

Kelsey, Candice M. *Generation MySpace: Helping Your Teen Survive Online Adolescence*. New York, NY: Marlowe & Company, 2007.

Levine, Madeline. *The Price of Privilege.* New York, NY: HarperColins Publishers, 2006.

Louv, Richard. *Last Child in the Woods*. Chapel Hill, NC: Algonquin Books of Chapel Hill, 2005.

Brock, Barbara. *Living Outside the Box: TV-Free Families Share Their Secrets*. Canada: Easter Washington University Press, 2007.

Rivkin, R. *The Great Outdoors: Restoring Children's Right to Play Outdoors*. Washington, D.C.: National Association for the Education of Young Children, 1995.

Shih, Clara. *The Facebook Era*. Boston, MA: Prentice Hall/Pearson Education, 2009.

Steyer, James P. *The Other Parent*. New York, NY: Atria Books, 2002.

Gordon, Gil. *Turn It Off: How to Unplug from the Anytime-Anywhere Office Without Disconnecting Your Career*. New York, NY: Three Rivers Press, 2001.

Winn, Marie. *The Plug-In Drug: Television, Computers, and Family Life*. New York, NY: Penguin Books 2002.

Conner, Bobbi. *Unplugged Play: No Batteries. No Plugs. Pure Fun.* New York, NY: Workman Publishing Company, 2007.

Krcmar, Marina. *Living Without the Screen: Causes and Consequences of Life Without Television.* New York, NY: Routledge, 2008.

Unplug Every Day: 365 Ways to Log Off and Live Better (Diary). San Francisco: Chronicle Books, 2014.

Worth, Pamela. Scandlyn, Tracy. *Unplug Your Family: Recipes for Creative and Meaningful Traditions*. Tiny Treks, 2013.

Adams, Gemini. *The Facebook Diet: 50 Funny Signs of Facebook Addiction and Ways to Unplug with a Digital Detox*. London, England: Live Consciously Publishing, 2013.

Whitney-Reiter, Nancy. *Unplugged: How to Disconnect from the Rat Race, Have an Existential Crisis, and Find Meaning and Fulfillment*. Boulder, CO: Sentiment Publications, 2008.

Lanza, Mike. Playborhood: *Turn Your Neighborhood into a Place for Play*. Menlo Park: Free Play Press, 2012.

Grossman, Elizabeth. *High Tech Trash: Digital Devices, Hidden Toxics, and Human Health*. Washington, D.C.: First Island Press, 2006.

Mander, Jerry. *Four Arguments for the Elimination of Television*. New York, NY: HarperColins, 1977.

Schor, Juliet B. *Born to Buy: The Commercialized Child and the New Consumer Culture*. New York, NY: Scribner, 2004.

Fielding, Orianna. *Unplugged: How to Live Mindfully in a Digital World*. London, England: Carlton Books Limited, 2014.

Kingsolver, Barbara. *Animal, Vegetable, Miracle: A Year of Food Life*. New York, NY: HarperColins Publishers, 2007.

Websites:

TV Turnoff Challenge: http://www.tvturnoff.net

The International Campaign Against Television: http://www.whitedot.org

Pew Research Center: http://www.pewinternet.org

Campaign for Commercial-Free Childhood: http://www.commercialfreechildhood.org/issues

National Day of Unplugging: http://nationaldayofunplugging.com

Unplug App: http://www.weareunplugged.com

Unplug & Reconnect: http://unplugreconnect.com

The Ultimate Guide to Unplugging: http://visual.ly/ultimate-guide-unplugging

Children & Nature Network: http://www.childrenandnature.org

Book Club Guide: The Questions

We entered into this experiment not knowing what the outcome would be, but there were some questions that we now have answers to. Invite your book club to answer these questions and discuss.

1. *How did doing away with technology affect our time, money, relationships and the environment?*
2. *Did we gain quality family time by turning the TV off? What did we do instead?*
3. *How did life without technology affect our five-year-old daughter? Did she protest? How did she change over the year? Do you think we were being 'mean' by depriving her of technology?*
4. *Does technology save time or waste time?*
5. *Is technology an expense we could live without or did it help us save money?*
6. *Does technology hinder relationships by limiting face-to-face communication? Or does it improve relationships by making it possible to reach friends easily and instantly?*
7. *Did a life without technology bring more intimacy to our marriage? Or did it cause distance by not being able to keep in touch constantly?*

8. *Did living a life without e-mail, social networking, and Internet access cause a strain on friendships due to going against the mainstream? Did we become social outcasts? How did we handle the peer pressure?*
9. *Does technology help the environment by using less paper? Or does it hurt the environment by wasting electricity?*
10. *How did we handle the inconveniences of not having the Internet to look up phone numbers, movie listings, get directions or order items online?*

Bonus Question: *What are some ways you can limit your technology usage? How do you think it will affect you, your friendships, marriage, and your children?*

Read More:

What to know even more about what happpended during the Kolberg's year uplugged? Read additional material that was cut from the book at **www.ayearunplugged.com**.

Other books by the author:

Six Seasons In Oz: 45 Weeks of Adventures in Sydney and *Beyond*

Building a Lasting Marriage: A Couple's Guide to Happily Ever After

23079255R00219

Printed in Poland
by Amazon Fulfillment
Poland Sp. z o.o., Wrocław